A LEVEL MATHE

MECHANICS 1

REVISION WORKBOOK

Version M

P. Thorns
J. King

Alpha Workbooks

Alpha Workbooks

Published by Alpha Workbooks.

Printed by *The Favil Press Ltd.,127 South Street, Lancing, Sussex BN15 8AS*

ISBN 1 903406 04 8

About This Book

Mechanics 1

This workbook is designed to be used as a revision tool as you prepare for your Module M1examination.
It consists of:

- *revision notes*
- *formulae*
- *worked examples*
- *examples for you to complete*
- *practice questions including exam type questions.*

We suggest you write in this book in pencil. There should be plenty of room in the example and answer boxes for your answers, but you can correct your work if necessary if you have used pencil. There are hints in the right hand column of some problems which you can cover up if you wish to complete the problems unaided.
There are mixed questions at the end of each section and suggested solutions in the answers section at the back of the book. At the end of the book there are some exam type questions.
Do not think you can work through this book in one attempt. It will be far more useful to attempt one section at a time and ensure you have all the rules and methods secure before attempting the next section.

You should use this workbook in conjunction with past module M1 papers and consult with your Mathematics tutor to revise effectively.

Contents

MODELLING

Simple Modelling:
Here is an example of a problem considered using a simple mathematical model.

Neil has a choice of doing his weekend shop at a supermarket 5 miles away or at his local shop within walking distance. In a simple model he may consider price difference in the two shops and travelling expenses for the Supermarket.
He knows that the Supermarket prices are 10% cheaper than the local shop, but that petrol and running costs of the car could be estimated at 50p per mile.
Neil now has to apply his model.
If he spent £10 in the local shop, what would be the cost in the Supermarket?
Prices 10% cheaper (£9) but add on travelling costs (5 × 50p) - total cost £11.50.
So a trip to the Supermarket is not worth it.
If Neil planned to spend £30 at the local shop, what would be the cost in the Supermarket?
Cost £27 + 5 × 50p = £29.50   now the trip may seem worth it.

Mathematical Models in Mechanics

(1) Take a real life problem	*Shall Neil use the Supermarket?*
(2) Make some simplifying assumptions to produce a model you can deal with (3) Present the problem in mathematical terms - that is produce a mathematical model	*Neil wants a simple model: he only considers 2 factors - price and travelling costs. He uses a general price comparison (assuming prices at the Supermarket are always 10% cheaper) and an estimate for travel costs per mile.*
(4) Solve this problem (either theoretically or experimentally) and produce results	*Neil tests his model with 2 possible shops - £10 and £30.*
(5) If you are satisfied with your solution you have finished, if not - go back to (2) and try improving your assumptions.	*Neil is happy that if he wants to spend £30 or more a trip to the Supermarket is worthwhile.*

In Mechanics there are many models to help explain the effects of forces on objects. In the modelling you use in the Mechanics 1 module you need to be aware of some frequently used vocabulary.

Basic Vocabulary for Modelling in Mechanics:

PARTICLE A particle is the name given to a body considered as a point in space, possessing mass but no size, (sometimes called a "point mass") if its dimensions are very small compared with the other lengths in the problem.

Many objects in problems in Mechanics are treated as particles even when they clearly have dimensions, to make the modelling easier. "Particle mechanics" avoids complications such as air resistance and rolling.

LIGHT An object is light if its mass is so small compared to the masses of other objects in the problem that its mass can be considered as zero.
If an object with mass is suspended by a string, the string is normally considered to be light and its mass is ignored.

SMOOTH A smooth surface is one which offers so little resistance to another body sliding over it that the friction is considered to be zero.
A smooth pulley is one which is considered to have no friction in its bearings. The tension in the string will be the same on either side of the pulley.

UNIFORM A body of uniform mass is one where equal amounts have equal masses.
Equal lengths of a uniform rod will have equal masses.
Equal areas of a uniform sheet of metal will have equal masses.

INEXTENSIBLE An inextensible (or inelastic) string is one whose length does not change even when acted upon by forces.

THIN A body is called thin when its thickness is very small compared to its width and length.
A piece of card or a sheet of metal would be considered thin. Each can be considered in a mechanics model as a two dimensional object.

RIGID A body is called rigid if it keeps its shape when acted on by forces.

Setting up simple models in mechanics

1. A slate sliding down a roof, inclined at 15° to the horizontal. Assumptions: The slate can be modelled as a particle The roof can be modelled as a smooth plane The slate moves in a vertical plane 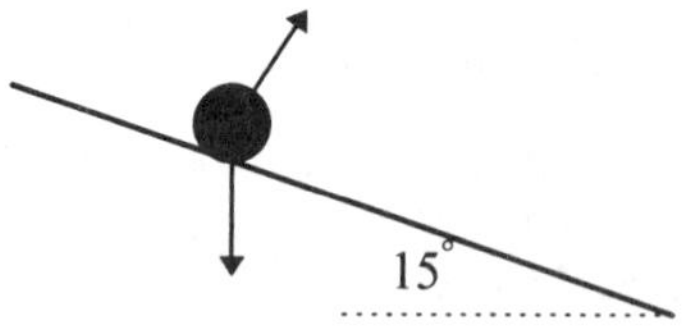 Once the forces have been added to the diagram, F = ma and the equations of motion with constant acceleration can be used	*You can draw a 2 dimensional diagram if the slate is moving in a vertical plane.*
2. A tennis ball in flight. Assumptions: The ball can be modelled as a particle No air resistance acts on the ball The ball moves in a vertical plane 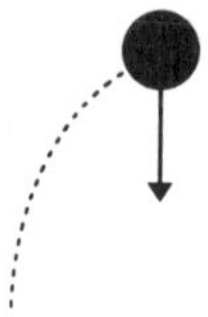 The only force acting on the ball is its weight. F = ma can then be used and the equations of motion with constant acceleration.	*With these assumptions you ignore the any spin on the ball or effect of turbulence*
3. A rope at 30° above the horizontal is pulling a boat along a horizontal chute. The rope passes over a pulley and is being pulled vertically downward with a force of 500 newtons. Assumptions:	*Try this one yourself.* *List the assumptions and then draw a diagram showing what forces are acting.*

VECTORS

Using vectors:

Vectors are quantities that have magnitude (size) and direction. Examples of vectors are displacement, velocity, acceleration and force.

If we consider vectors acting in a straight line:

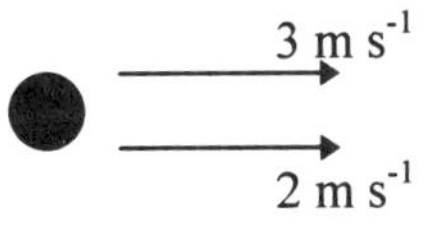

Consider a particle moving at 3 m s^{-1} East on a track which is moving at 2 m s^{-1} East. The particle would move 3 m in 1 second and 2 m in the same second. This is equivalent to moving 5 m in the second i.e. a velocity of 5 m s^{-1} East.

We can represent any vector by a straight line with its length equal to the magnitude of the vector and the direction of the line representing the direction of the vector.

3 m s^{-1} 'together with' 2 m s^{-1} would be drawn.

'followed by' The Resultant is the length and direction of the resulting line. 5 m s^{-1} East

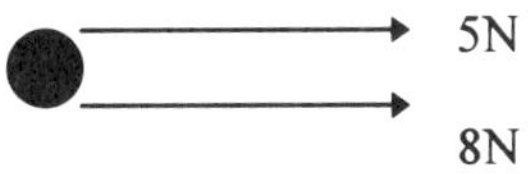

A 5 N force 'together with' a 8 N force acting on a particle along a line. This is equivalent to a force of 13 N acting in the same direction.

We can represent any vector by a straight line with its length equal to the magnitude of the vector and the direction of the line representing the direction of the vector.

'followed by'

The Resultant is the length and direction of the resulting line. 13 N in the same direction.

For vectors in 2 dimensions we just extend this idea.

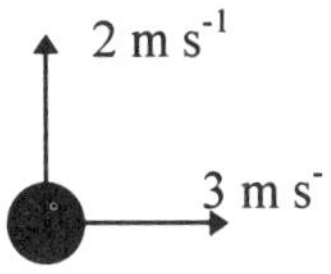

Consider a particle moving at 3 m s^{-1} East on a surface that is moving 2 m s^{-1} North. In 1 second the particle would move 3 m East and 2 m North.

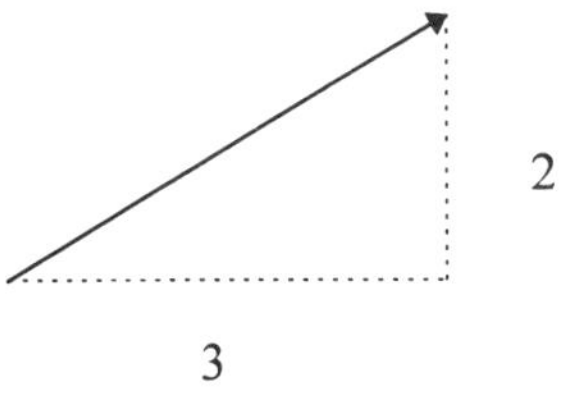

3 m East 'followed by' 2 m North.
The resultant velocity is the length and direction of the resulting line. (i.e. in one second the particle will move $\sqrt{13}$ m at an angle 33.69° to the East direction). The resultant velocity is $\sqrt{13}$ m s^{-1} acting at an angle 33.69° to the East direction.
This can be obtained by calculation (or scale drawing - but do not do a scale drawing unless told to do so).

and

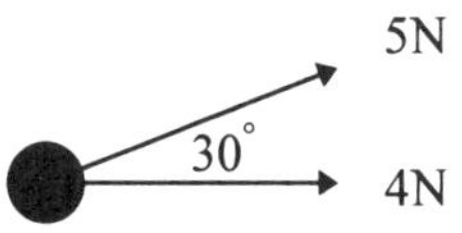

A force 4 N acting on a particle in the x direction 'together with' a force of 5 N acting on a particle at 30° to the x direction.

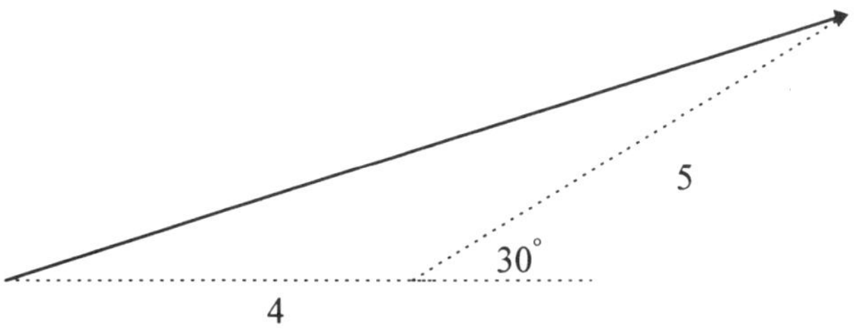

4 in the x direction 'followed by' 5 at 30° to the x direction is equivalent to 8.70 N at 20.5° to the x direction.
This can be obtained by calculation (or scale drawing - but do not do a scale drawing unless told to do so).

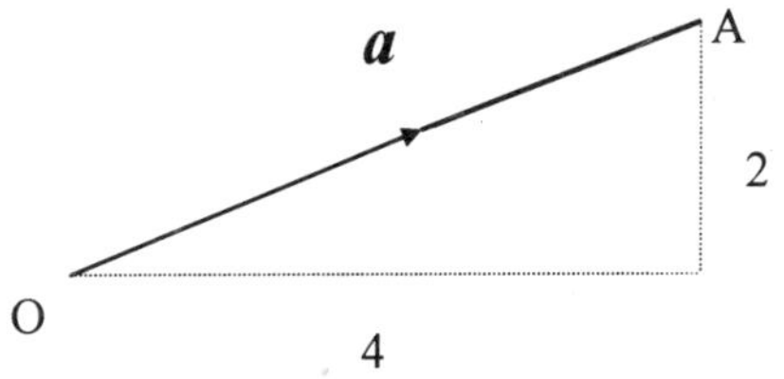

Vectors can be represented in various forms. (Notice the bold type used in books and exams.)

Here $\boldsymbol{a} = \underline{a} = \overrightarrow{AB} = 4\mathbf{i} + 2\mathbf{j} = \begin{bmatrix} 4 \\ 2 \end{bmatrix}$

The magnitude of the vector is written

$|\boldsymbol{a}|$ or $|\overrightarrow{AB}| = \sqrt{4^2 + 2^2}$ or just a (not in bold).

If $\underline{a} = \begin{bmatrix} 3 \\ -4 \end{bmatrix}$ $\underline{b} = \begin{bmatrix} 5 \\ 1 \end{bmatrix}$ $\underline{c} = \begin{bmatrix} 0 \\ -2 \end{bmatrix}$

> These vectors could be forces in Newtons or velocities in m s^{-1} or acceleration in m s^{-2} (As shown before)

show how the following vectors are obtained graphically on the grid below. $\underline{a} - 2\underline{b}$ is shown. (the resulting vector is called the *resultant*):

$2\underline{b}$ $\underline{a} - \underline{b}$ $2(\underline{a} + \underline{c})$ $2\underline{a} + \underline{b} - 3\underline{c}$

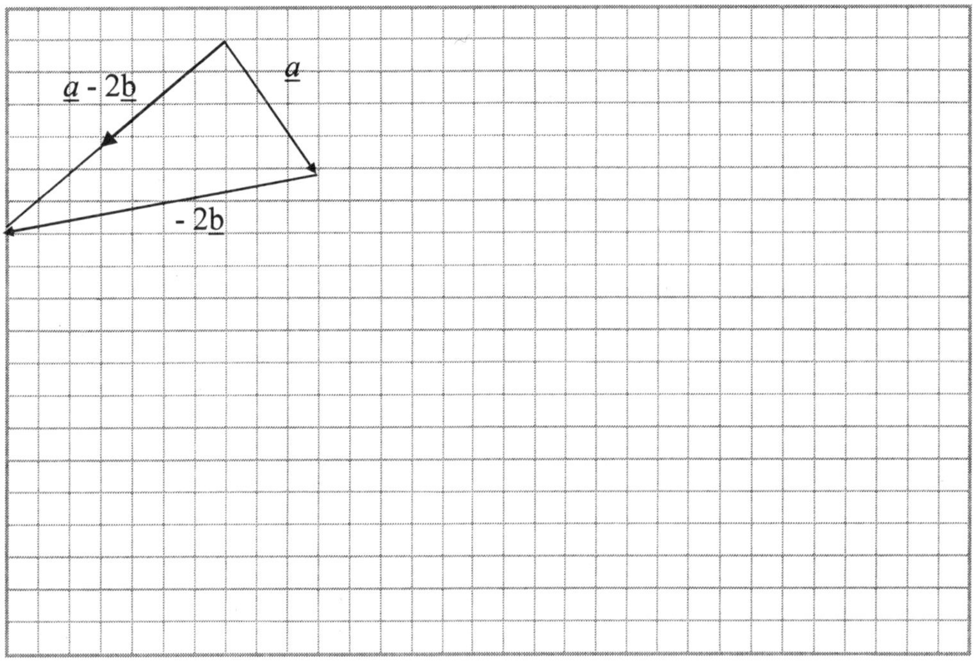

> Think: $\underline{a} + (-2\underline{b})$
> $\underline{a}$ followed by $-2\underline{b}$

Now work out the *resultant* vectors numerically (work out your answers in column vector and **i**, **j** form)

$2\underline{b} = 2 \times \begin{bmatrix} \quad \end{bmatrix} = \begin{bmatrix} \quad \end{bmatrix}$ $2\underline{b} = 2(5\mathbf{i} + \mathbf{j}) =$

$\underline{a} - \underline{b} = \begin{bmatrix} \quad \end{bmatrix} - \begin{bmatrix} \quad \end{bmatrix} = \begin{bmatrix} \quad \end{bmatrix}$ $\underline{a} - \underline{b} =$

$2(\underline{a} + \underline{c}) =$ $2(\underline{a} + \underline{c}) =$

$2\underline{a} + \underline{b} - 3\underline{c} =$ $2\underline{a} + \underline{b} - 3\underline{c} =$

> When finding the magnitude and direction of a resultant vector always sketch the vector. Here the sketches are on the grid.

Find the magnitude and direction of the following: $2\underline{b}$, $\underline{a} - \underline{b}$

Magnitude, $|2\underline{b}| =$ $\tan\alpha =$ direction = to the **i** direction

Magnitude, $|\underline{a} - \underline{b}| =$ $\tan\alpha =$ direction =

For $\underline{a} - \underline{b}$ just use the triangle 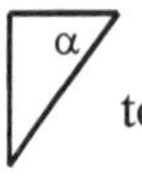 to work out the angle and then subtract from 180° to find the angle with the **i** direction.

In three dimensions

If $\boldsymbol{a} = \begin{bmatrix} 2 \\ -3 \\ 1 \end{bmatrix}$ then also $\boldsymbol{a} = 2\mathbf{i} - 3\mathbf{j} + \mathbf{k}$ and $|\boldsymbol{a}| = \sqrt{2^2 + (-3)^2 + 1^2} = \sqrt{14} = 3.74$

When drawing vector diagrams or working out vectors from diagrams you may find it useful to use the following idea

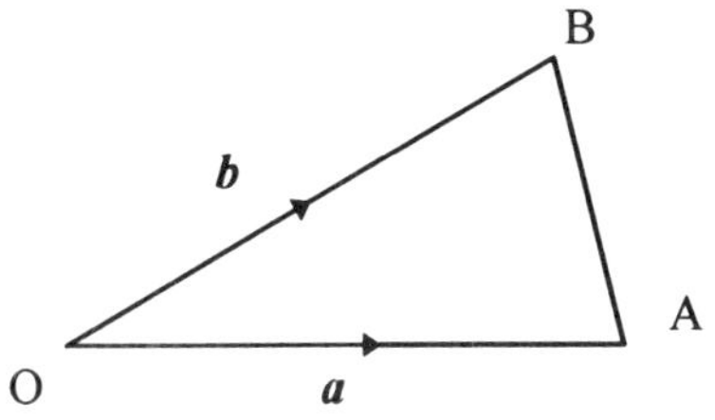

$\overrightarrow{AB} = \overrightarrow{AO} + \overrightarrow{OB}$

Think: 'is equivalent to' (=)

Think: 'followed by' (+)

So:

$\overrightarrow{AB} = -\boldsymbol{a} + \boldsymbol{b}$

$= \boldsymbol{b} - \boldsymbol{a}$

Position vectors:
These are vectors referred to an origin. So the position vector of the point (3,2) is $3\mathbf{i} + 2\,\mathbf{j}$.

Unit vectors:
To find a unit vector in any direction, just take any vector in that direction and divide by the magnitude of the vector.

To find a unit vector in the direction of the vector $5\mathbf{i} + 12\mathbf{j}$

Magnitude $\sqrt{5^2 + 12^2} = 13$

Unit vector is $\frac{5}{13}\mathbf{i} + \frac{12}{13}\mathbf{j}$

COMPLETE: OPRQ is a parallelogram with $\overrightarrow{OP} = \mathbf{p}$ and $\overrightarrow{OQ} = \mathbf{q}$ and M and N are the mid points of PQ and QR respectively. Find in terms of **p** and **q** (i) $\overrightarrow{PQ}$ (ii) $\overrightarrow{PM}$ (iii) $\overrightarrow{OM}$ (iv) $\overrightarrow{MQ}$ (v) $\overrightarrow{MN}$

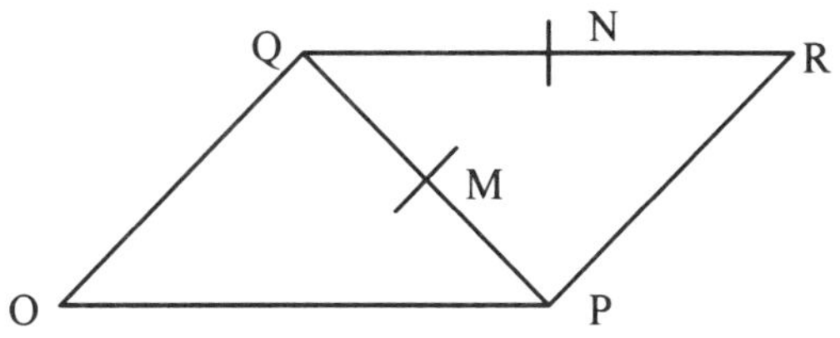

(i) $\overrightarrow{PQ}$ =	*$\overrightarrow{PO}$ 'followed by' $\overrightarrow{OQ}$.*
(ii) $\overrightarrow{PM}$ =	*$\frac{1}{2}$ of $\overrightarrow{PQ}$.*
(iii) $\overrightarrow{OM}$ =	*$\overrightarrow{OP}$ 'followed by' $\overrightarrow{PM}$.*
(iv) $\overrightarrow{MQ}$ =	*The same as $\overrightarrow{PM}$.*
(v) $\overrightarrow{MN}$ =	*$\overrightarrow{MQ}$ 'followed by' $\overrightarrow{QN}$.*

COMPLETE: Write down the position vectors of the points P(5,2), Q(7,1) and R(11,-1). Find the vectors $\overrightarrow{PQ}$ and $\overrightarrow{PR}$ and show that P, Q and R are on a straight line.

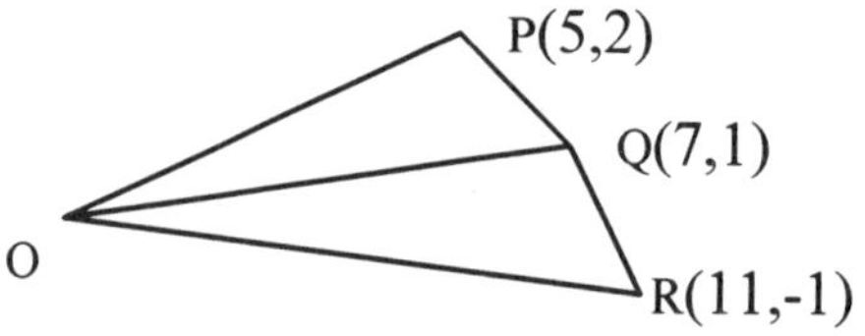

Position vectors

$\overrightarrow{OP} =$

$\overrightarrow{OQ} =$

$\overrightarrow{OR} =$

$\overrightarrow{PQ} = \overrightarrow{PO} + \overrightarrow{OQ} \quad =$

$\overrightarrow{PR} = \qquad\qquad =$

A sketch helps you to see what is going on.

Straight line if $\overrightarrow{PR}$ is a multiple of $\overrightarrow{PQ}$.

1. If $\boldsymbol{a} = \begin{bmatrix} 2 \\ 1 \end{bmatrix}$, $\boldsymbol{b} = \begin{bmatrix} -2 \\ 3 \end{bmatrix}$ and $\boldsymbol{c} = \begin{bmatrix} 5 \\ -1 \end{bmatrix}$ calculate

   (i) $\underline{a} + 2\underline{b}$

   (ii) $2\underline{a} + \underline{b} - \underline{c}$

   (iii) $\underline{a} + \underline{b} + \underline{c}$

   (iv) $3(\underline{a} - \underline{b} - \underline{c})$

This is how you would write the vectors(underlined).

2. Two forces of magnitude 6 N and 10 N act at an angle of 90° to each other. Calculate the magnitude and direction of the resultant of these two forces.

Answers to 3 sig. fig.

Draw one vector , 6 N, 'followed by' the other vector 10 N at 90°.

Give the direction to one of the vectors.

3. Two forces P and Q are of magnitude 12 and 8 newtons. Calculate the magnitude and direction of the resultant of P and Q if:
   (i) P and Q act in the same direction and are parallel

   (ii) P and Q act in opposite directions and are parallel

   (iii) P and Q are inclined at 120° to each other.

In each case sketch the vector diagram - using P 'followed by' Q.

First sketch P and Q at 120°. Then sketch P 'followed by' Q

Answers to 3 sig. fig.

4. Three vectors **p**, **q** and **r** are 3units due S, 4 units SW and 5 units due E. find by calculation the magnitude and direction of:	*Answers to 3 sig. fig.*
(a) **p** + **q**	*Draw a vector diagram.* ***P*** *'followed by'* ***q****.*
(b) **p** + **q** - 2**r**	***p*** *'followed by'* ***q*** *'followed by' -2****r****.*
5. A boat, whose speed in still water is 4 m s^{-1}, is heading directly across a river which is 200 m wide and flows at 3 m s^{-1}. Calculate	*Answers to 3 sig. fig.*
(i) the resultant speed and direction of the boat	*Draw a velocity vector diagram to find the resultant velocity. One vector 'followed by' the other vector.*
(ii) the time taken to cross the river	*Sketch the river and the resultant velocity vector. Use the angle to work out the distance to be covered(use a triangle) and hence the time from the speed.*
(iii) the distance downstream which it will be carried.	*Again just use the triangle.*

Resolving vectors:

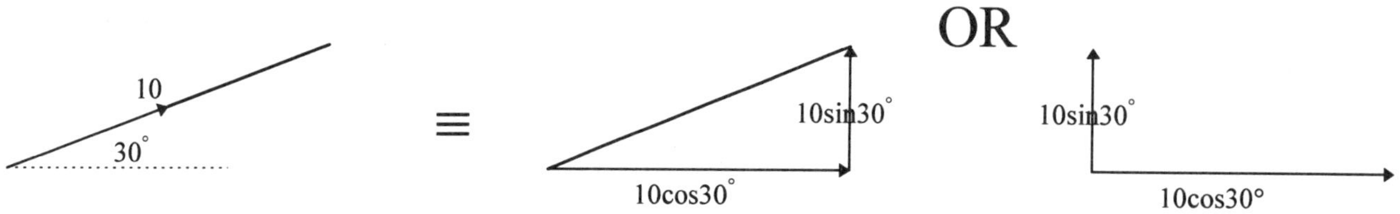

Remember: this vector could be a velocity or an acceleration or a force

So the vector can be written $10\cos30^\circ\mathbf{i} + 10\sin30^\circ\mathbf{j}$

These are called the components

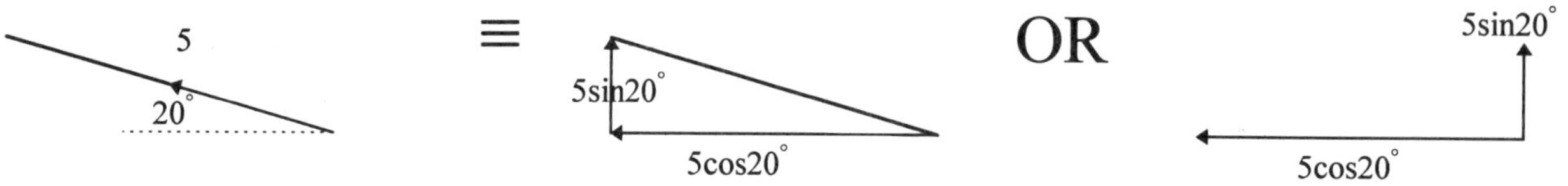

So the vector can be written $-5\cos20^\circ\mathbf{i} + 5\sin20^\circ\mathbf{j}$

When solving vector problems you can use the resolved vectors to make the calculations easier, (If you have three or more velocity vectors it would be easier to use the components of each vector and sum them rather than to draw triangles and use cosine/sine rules to find the resultant)

1. COMPLETE: A boat whose speed in still water is 2 m s^{-1} is heading on a bearing of 030° is being taken off course by a current of 1.5 m s^{-1} from 315° and a wind of 1 m s^{-1} (the effect the wind has on the boat) from 200°. Find the speed and direction of the boat.

Drawing one vector 'followed by' another 'followed by' another gives the following . You can see that it would not be easy to find the resultant vector.

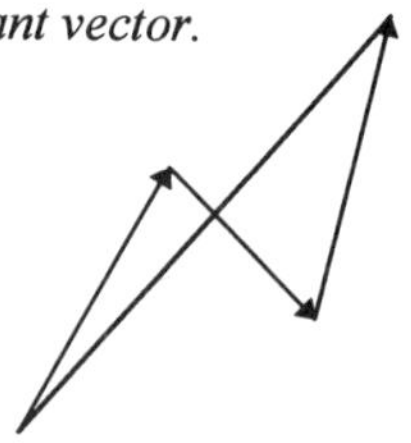

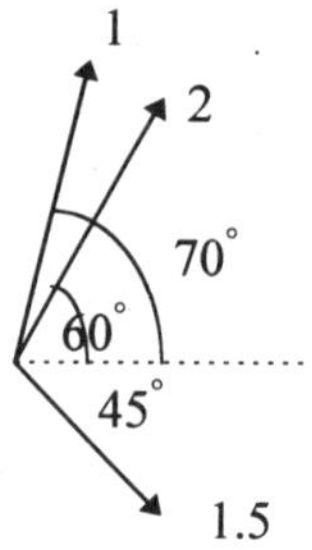

Resolving each vector:

Boat: $2\cos60^\circ\mathbf{i}$ + $\mathbf{j}$

Current: $\mathbf{i}$ - $\mathbf{j}$

Wind: $\mathbf{i}$ + $\mathbf{j}$

(The angles being worked out from the bearings.)

Resultant velocity:

$0.5\,\mathbf{i}$ + $\mathbf{j}$ +

Add the vectors.
Use decimals to at least 5 decimal places.

= $\mathbf{i}$ + $\mathbf{j}$

Collect together.

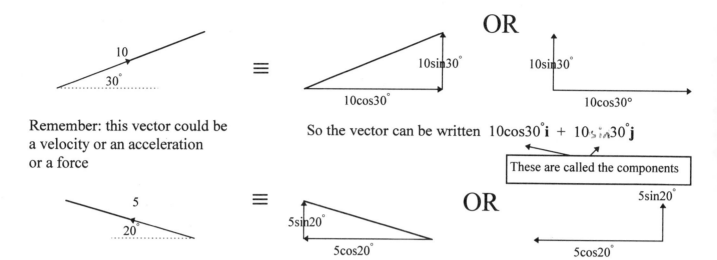

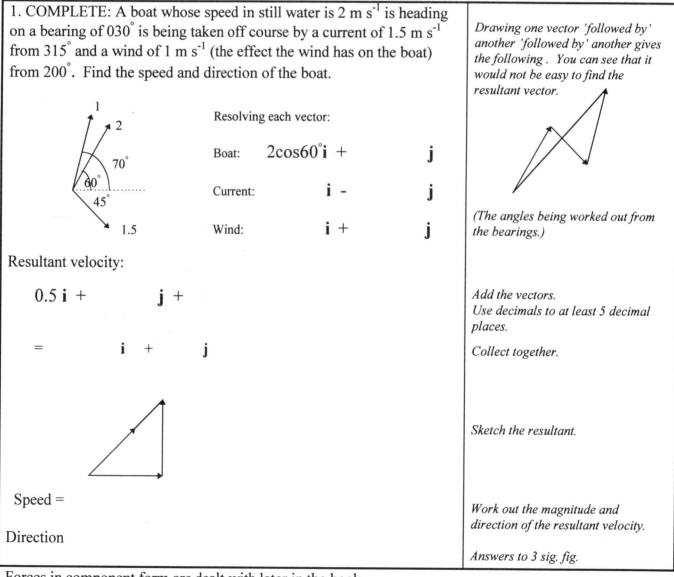

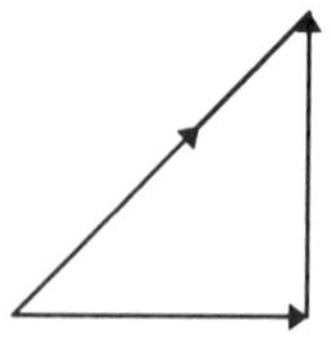

Sketch the resultant.

Speed =

Direction

Work out the magnitude and direction of the resultant velocity.

Answers to 3 sig. fig.

Forces in component form are dealt with later in the book.

MOTION IN ONE DIMENSION

Displacement is the change in position. It is a vector and therefore has magnitude and direction, In straight line motion an object can be forwards or backwards from its initial position and therefore a +ve or -ve value will show the direction.

If an object is travelling with constant speed,

$$speed = \frac{distance}{time}$$

If speed is not constant,

$$average\ speed = \frac{total\ distance}{total\ time}$$

Speed is a scalar quantity
Velocity is a vector quantity and therefore has magnitude (speed) and direction. In straight line motion the velocity can be in one of two directions i.e. +ve or -ve.

$$velocity = \frac{displacement}{time}$$

$$average\ velocity = \frac{total\ displacement}{total\ time}$$

Velocity is the rate at which the position of an object changes
e.g. a body changes from position 3 m to position 11 m in 2 seconds. Its change in position is 8 m in 2 seconds i.e. average velocity +4 m s^{-1} (4 metres per second in the +ve direction)

1. Complete: A woman walks 300 m due East in 100 secs. then 200 m due West in 80 secs Calculate her average speed and her average velocity.	
Diagram 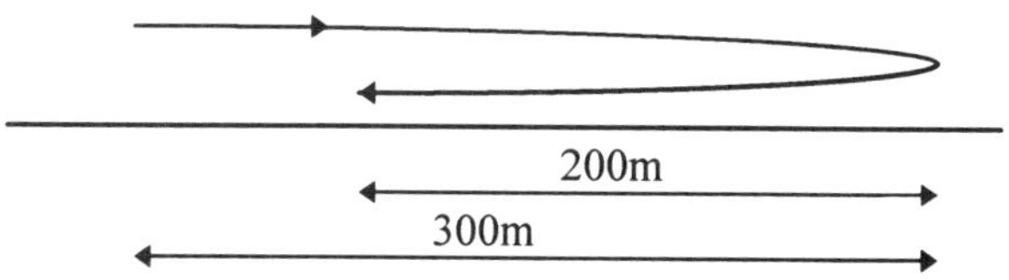	*Take East as the positive direction*
Total distance travelled =	*Remember distance just means ground covered in this case.*
Total displacement =	*Displacement means position relative to starting point.*
Average speed = =	*Just use the formulae above*
Average velocity = =	*Don't forget to state the units used and the direction*

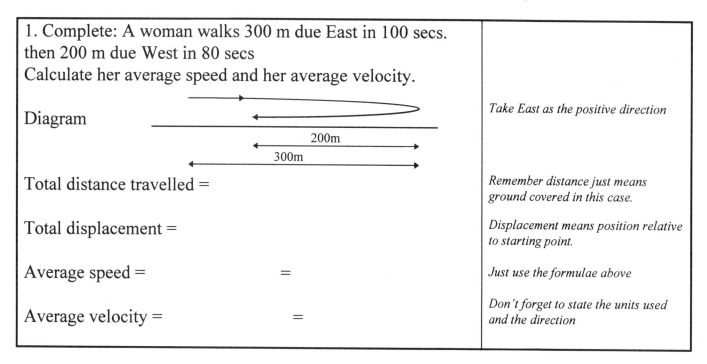

2.

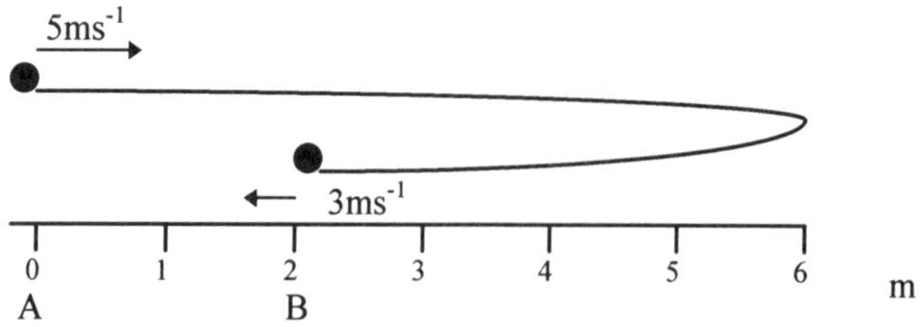

Take 'right as the positive direction.

A body at A moves as shown to B

Distance travelled from A to B =

Displacement of B from A =

Indicate the direction by + or - sign for displacement and velocity.

Speed of body at A =

Velocity of body at A =

Speed of body at B =

Velocity of body at B =

3. Convert 40 km h^{-1} to m s^{-1} .

First change the km to m.

Then change hours to sec.

4. Convert 4 m s^{-1} to km h^{-1} .

First change the sec to hours.

Then change m to km..

5. P, Q and R are three points, in that order, on a straight road with PQ = 6 km and QR = 8 km. A man travels from P to Q at 12 km h^{-1} and from Q to R at 20 km h^{-1}. Calculate
   (i) the time taken to travel from P to Q
   (ii) the time taken to travel from Q to R
   (iii) the average speed of the man in travelling from P to R

You can use

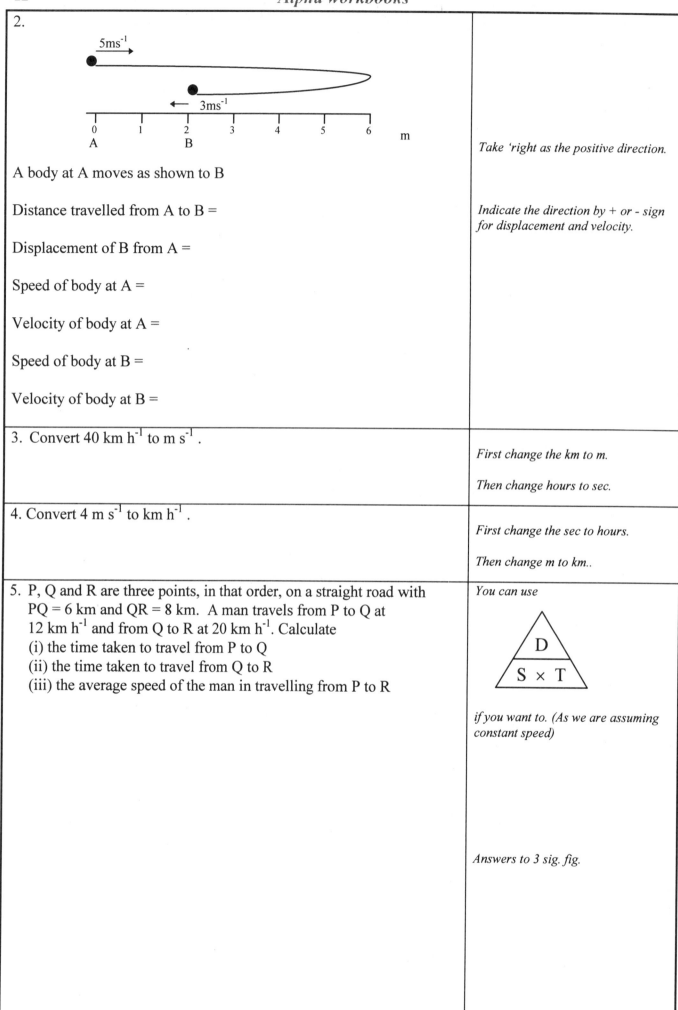

if you want to. (As we are assuming constant speed)

Answers to 3 sig. fig.

Distance/position time graphs:

Gill walks at constant speed to her friend's house 120m away in 60seconds. After talking to her friend for 30seconds she returns home at constant speed in 40seconds.

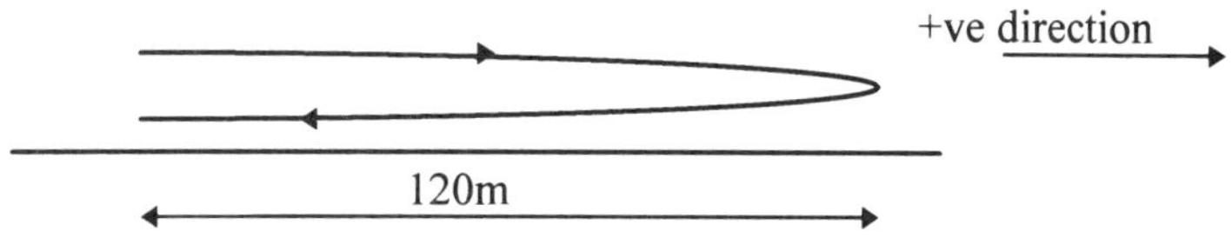

Distance time graph

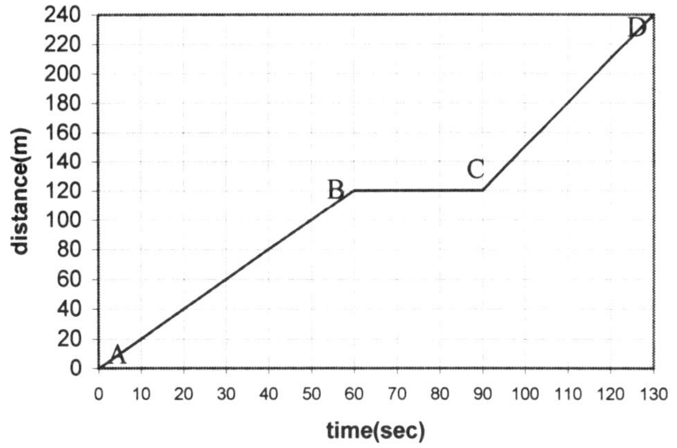

A to B 120m in 60sec
i.e. speed $2ms^{-1}$

Equal to the gradient from A to B

B to C waiting (zero speed zero gradient)

C to D 120m in 40sec
i.e. speed $3ms^{-1}$

Equal to the gradient from C to D

Position time graph (if the position is related to a fixed point or origin this can also be called a displacement time graph)

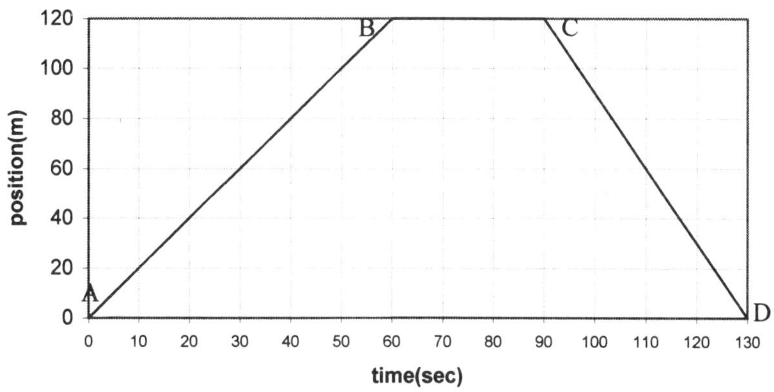

A to B 120m in 60sec
i.e. velocity $2ms^{-1}$

Equal to the gradient from A to B

B to C waiting (zero velocity zero gradient)

C to D 120m in 40sec
i.e. velocity $-3ms^{-1}$ (negative indicates velocity in the opposite direction)

Equal to the gradient from C to D

The gradient of a position time graph is equal to the velocity.

1. A footballer is doing some fitness training. Starting from A he runs at constant velocity to B in 4 seconds. He then immediately turns around and runs at constant velocity to C in 8 seconds. Sketch the position time graph.

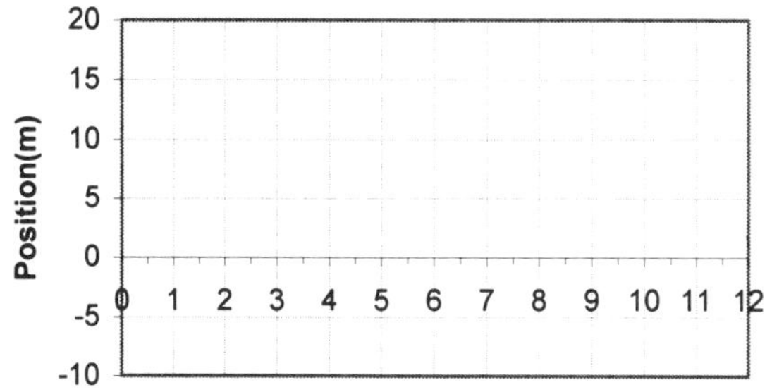

Work out the velocities for the two parts of the motion.

One will be negative.

2. A goods truck is shunted 50 m forward in 10 seconds, then 28 m back in 7 seconds then 36 m forward in 12 seconds.
(i) Sketch the distance-time graph
(ii) Calculate the speeds for each section
(iii) Find the average speed for the whole journey
(iv) Sketch the position-time graph
(v) Calculate the velocity for each section
(vi) Find the average velocity for the whole journey

Assumption: Constant speeds for each part of the motion

Speed = gradient on dist/time graph

A sketch here may help 'to see' what is going on

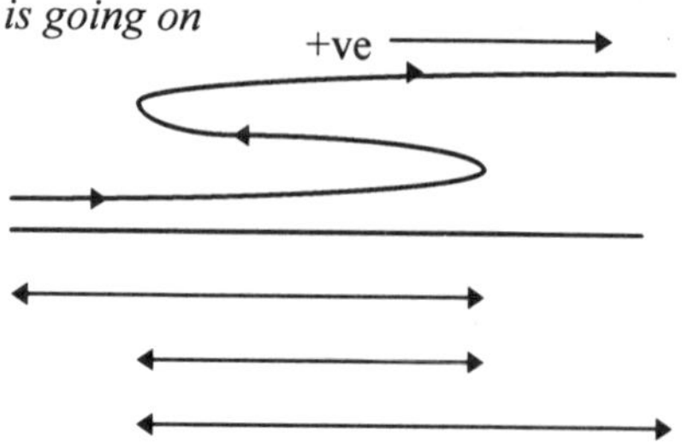

put in the distances.
Sketch the graph. Use a ruler.

(i) COMPLETE:

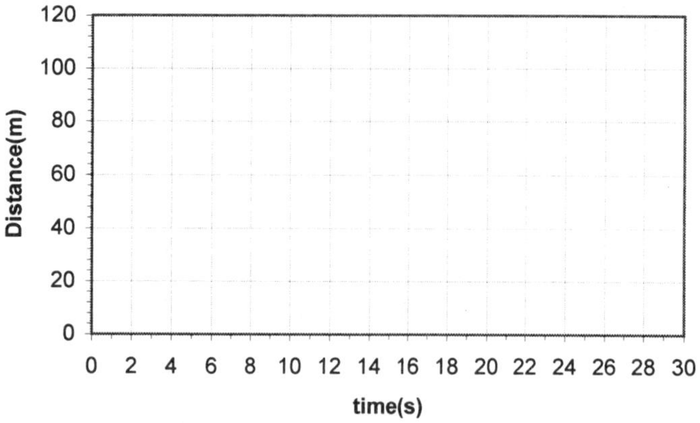

(ii) $speed = \dfrac{distance}{time}$ so need the gradient on the graph

First part of motion Speed = = ms^{-1}

Second part of motion Speed = = ms^{-1}

Last part of motion Speed = = ms^{-1}

Gradient of graph. Use 50 / 10 as these numbers are given.

(iii) Average speed = $\dfrac{total\ dist.}{total\ time}$ =

(iv)

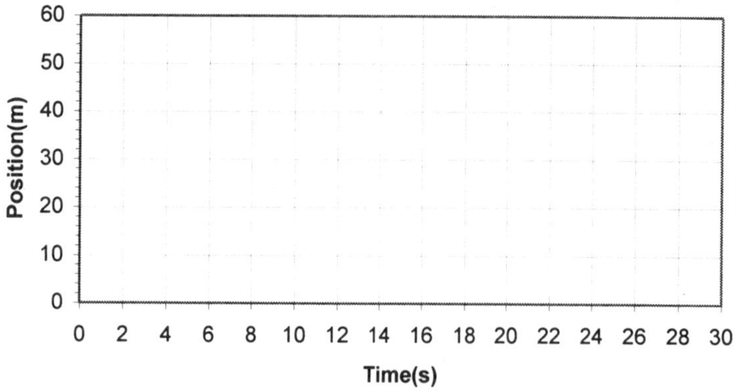

Sketch the graph. Use a ruler.

(v) First part of motion velocity = = ms^{-1}

Second part of motion velocity = = ms^{-1}

Last part of motion velocity = = ms^{-1}

One of these will be negative.

(vi) Average velocity =

Velocity time graphs:

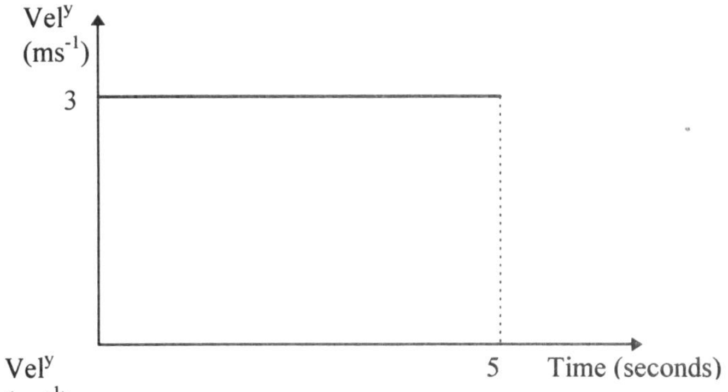

This graph shows constant velocity of 3 m s^{-1} for 5 seconds.
The body travels $3 \times 5 = 15$ m.
This is also the area under the graph.

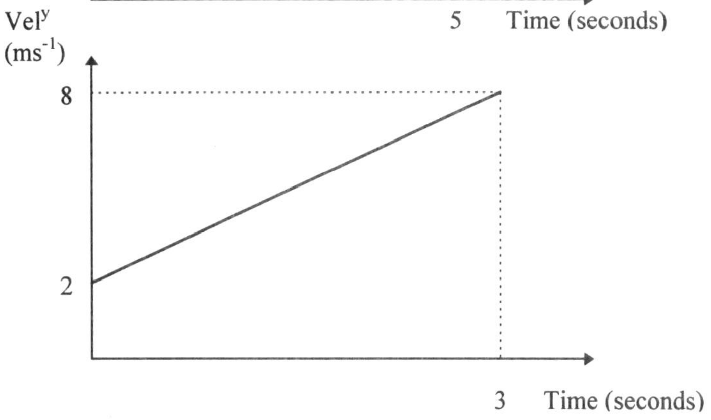

Acceleration is the rate at which the velocity changes.
Here the velocity changes from 2 m s^{-1} to 8ms^{-1} in three seconds.
The velocity changes by 6 m s^{-1} in 3 seconds which is $\frac{6}{3} = 2$ m s^{-1}per sec i.e. the acceleration is 2 m s^{-2}.
This is the gradient of the line.

The area between a velocity time graph and the time axis is equal to the distance. For *distance* add the areas. For *displacement* add the areas above the time axis and subtract the areas below the time axis.

The gradient of a velocity time graph is equal to the acceleration.

Complete: A particle starting from rest accelerates uniformly in a straight line for 4 seconds by which time it is travelling at 14 m s^{-1}. It then travels at this velocity for 3 seconds before uniformly returning to rest in 5 seconds.

a) draw the velocity time graph for this particle.

Only a sketch. No need for a scale. Use a ruler and mark the axes at the important points.

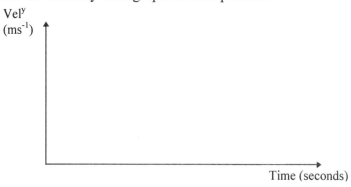

For part b) we could use the equations of motion - but using the graph is easier.

b) find the accelerations of the particle and the distance travelled.

Constant acceleration of — = m s^{-2} for 4 seconds.

Gradient of graph.

Constant acceleration of - — = - m s^{-2} for 5 seconds.

Gradient of graph. (Note - negative i.e. negative acceleration which is deceleration)

Distance travelled = = m

Area under graph. (Area of a trapezium or two triangles and a rectangle.)

1. This velocity time graph shows a body which accelerates uniformly for 5 seconds after which time it maintains its final velocity for 3 seconds.
   Find (i) the initial velocity, u
   (ii) the final velocity, v
   (iii) the acceleration , a, during the first 5 seconds
   (iv) the distance travelled by the body during the 8 seconds.

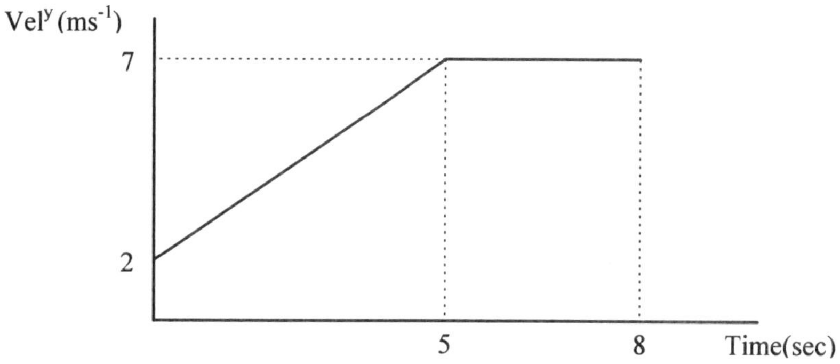

u =

v =

a =

Gradient of the graph.

distance =

Area between graph and the t axis.

2. This velocity time graph shows a body moving at constant velocity for 4 seconds and then decelerating uniformly for 6 seconds. Find
   (i) the initial velocity, u
   (ii) the final velocity, v
   (iii) the acceleration , a, during the last 6 seconds
   (iv) the displacement from its starting point
   (v) the total distance travelled by the body.

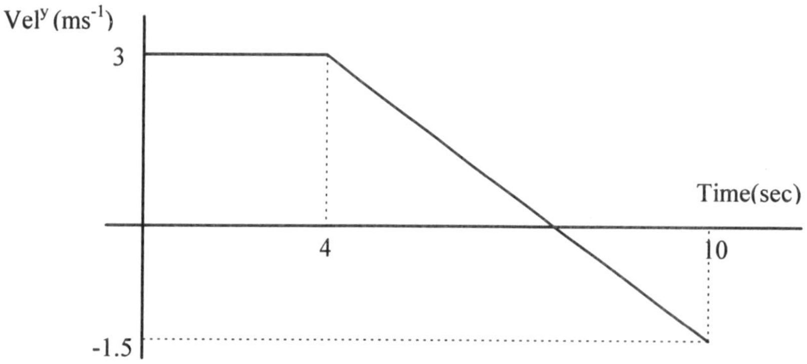

u =

v =

a =

Negative.

displacement =

To find the distance and displacement you will need to work out where the line cuts the time axis. Use areas.
For displacement: above the time axis is +ve and below is -ve. Then just add.

distance =

For distance: make both the areas +ve and add.

3. A particle travelling in a straight line has an initial velocity of $10\ \text{m s}^{-1}$ and an acceleration of $-2\ \text{m s}^{-2}$.
   (a) Draw a velocity time graph for the first 7 seconds of motion.
   (b) Find the displacement from the start after 7 seconds.
   (c) Find the distance travelled in 7 seconds.

Only a sketch. No need for a scale. Use a ruler and mark the axes at the important points

Displacement =

Use areas. Above the time axis is +ve and below is -ve. Then just add.

Distance travelled =

Make both the areas +ve and add.

4. A cyclist rides along a straight road from P to Q. He starts from rest at P and accelerates uniformly to reach a speed of $15\ \text{m s}^{-1}$ in 10 secs. He maintains this speed for 20 seconds, then slows down uniformly to stop at Q. The whole journey takes 36 secs.
   (i) Draw a velocity-time graph for the journey
   (ii) Calculate the acceleration for the first part of the journey
   (iii) Calculate the deceleration for the last part of the journey
   (iv) Find the total distance travelled by the cyclist.

Just a sketch. Use a ruler but do not worry about scales. Label important points on the axes.

5. The position of a particle moving along a straight line is given by $s = t^2 - 5t + 3$.
   (i) Draw a diagram to show the path (position) of the particle over the first 6 seconds of its motion.
   (ii) How far does the particle travel in the first 6 seconds?
   (iii) Find the average speed over the first 6 seconds
   (iv) Find the average velocity over the first 6 seconds.

6. Two cars Z and X start from rest, travel equal distances in the same time and finish at rest. Car Z accelerates at 1.25 m s^{-2} for 16 seconds, travels at constant speed for 20 seconds and is then brought to rest with a uniform deceleration of 2 m s^{-2}. Car X accelerates uniformly at x m s^{-2} and is then brought to rest with a uniform deceleration of $3x$ m s^{-2}. Sketch the velocity-time graphs of the two cars on the same diagram and calculate:
   (i) the greatest speed reached by Z
   (ii) the total distance travelled by Z
   (iii) the time for which X accelerates
   (iv) the greatest speed reached by X

Motion in a straight line with constant acceleration:

The five possible equations of motion for constant acceleration are as follows:

$$v = u + at$$

$$s = ut + \tfrac{1}{2}at^2$$

$$v^2 = u^2 + 2as$$

$$s = \tfrac{1}{2}t(u + v)$$

$$s = vt - \tfrac{1}{2}at^2$$

Where
s = displacement
u = initial velocity
v = final velocity
a = uniform acceleration
t = length of time interval

You may only be used to using the first 3 or 4 of these. Just use the ones you recognise.

1. A body is moving in a straight line from P to Q

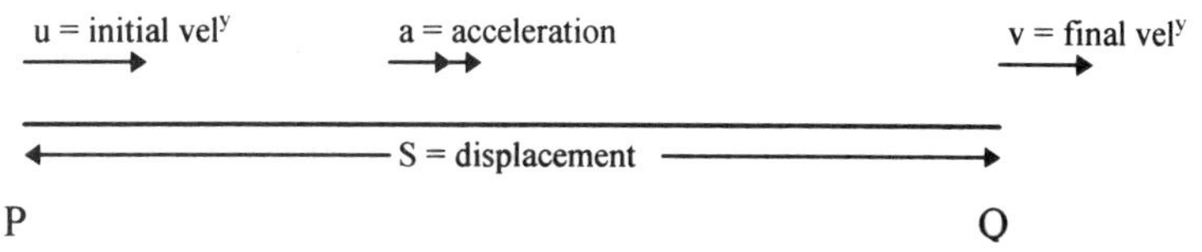

If:

a) $u = 0$, $a = 7$, $t = 8$, find s

b) $u = 5$, $a = 3$, $t = 5$, find v

c) initially at rest, $a = 2$, $t = 5$, find s

d) $u = -5$, $v = 7$, $t = 10$, find s

e) $u = 7$, $v = 9$, $s = 2$, find a

f) $v = 32$, $a = 6$, $t = 3$, find the initial velocity

g) $u = 17$, $v = -13$, $s = 10$, find a

h) $s = 32$, $a = 2$, $t = 4$, find u

i) $s = 144$, $t = 6$, $a = -2$, find v

j) $u = 5$, $t = 10$, $a = -3$, find s

k) $u = 6$, $v = -2$, $t = 6$, find a

Careful in all questions.
$speed = \frac{distance}{time}$ *is for zero acceleration. (constant velocity) Do not use it for these questions.*

2. A dog accelerates from rest to 15 m s^{-1} in a distance of 30 m. Find its acceleration (assume constant).	*Write down u = v = s = and a = ?* *Write down the equation you will use. Then use it to find a.*
3. A car decelerates uniformly from 20 m s^{-1} until it stops after 6 seconds. What is its deceleration?	*Write down u = v = s = and a = ?* *Write down the equation you will use. Then use it to find a.*
4. A car accelerates uniformly from rest to 96 km h^{-1} (about 60 mph) in 12 seconds. Find its acceleration and the distance travelled during acceleration	*Change to m s^{-1}.* *Write down u = v = t = and a = ?* *Write down the equation you will use. Then use it to find a.* *Then find s.*
5. An electric hare travelling on a straight track at 5 m s^{-1} passes a stationary greyhound. One second later the greyhound sets off after the hare, accelerating at 4 m s^{-2}. How long is it before the greyhound catches up with the hare and how far has the greyhound travelled?	*If the hare has been moving for t seconds since passing the greyhound then the greyhound has been moving one second less i.e. for (t-1) seconds.* *Use '$s = ut + \frac{1}{2}at^2$' for both (with (t-1) for t for the greyhound) and equate.*
6. A boy on a sledge pushes himself off at the top of a slope with a speed of 1.5 m s^{-1}. He moves with constant acceleration and after 10 seconds is moving at 4 m s^{-1}. (i) What is the acceleration of the boy and sledge? (ii) The length of the slope is 200 m. What is the speed of the boy and sledge at the bottom of the slope?	*Sketch. Show the initial speed, the distance, final speed and the acceleration.* *Use the constant acceleration equations.*

Acceleration graphs:

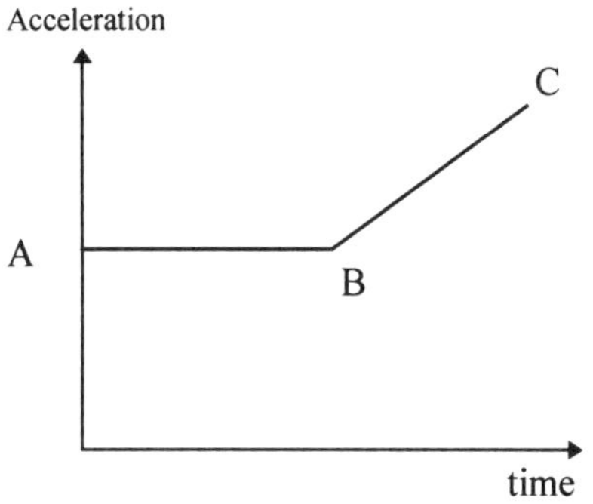

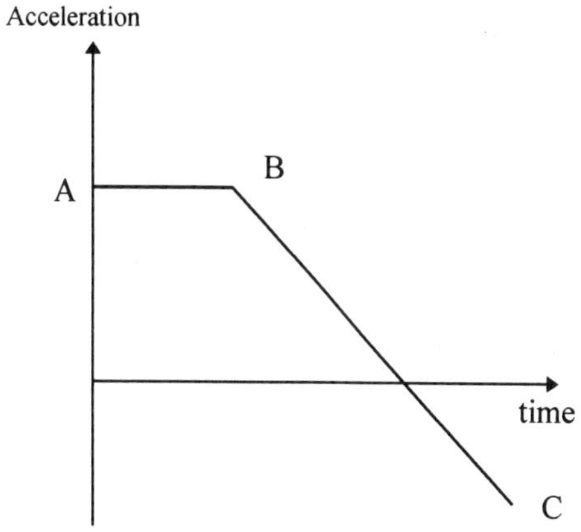

AB shows constant acceleration. You can use the constant acceleration formulae.

BC shows variable acceleration. *A* is a function of t. You can use integration to find velocity and or displacement.(See chapter on general motion of a particle.)

<u>OR</u>

Area under graph = increase in velocity

Above t axis +ve
Below t axis -ve (i.e. decrease in velocity)

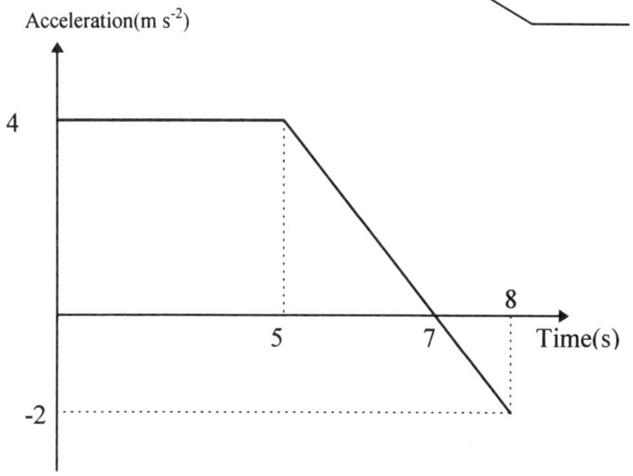

The acceleration of a particle is shown. If the particle has initial velocity of 3 m s^{-1} find the velocity after (a) 5 second (b) 10seconds.

(a) Area = 4×5 = 20 = increase in velocity
Velocty after 5 seconds = 3 + 20 = 23 m s^{-1}.

(b) Area above = .5 × 2 × 4 = 4
Area below = .5 × 1 × 2 = 1

velocity after 10 seconds = 23 + 4 - 1 = 26 m s^{-1}

1. A particle moves with the acceleration shown. If its initial velocity is 2 m s^{-1}, find its velocity after (a) 2 seconds (b) 5 seconds (c) 8seconds.

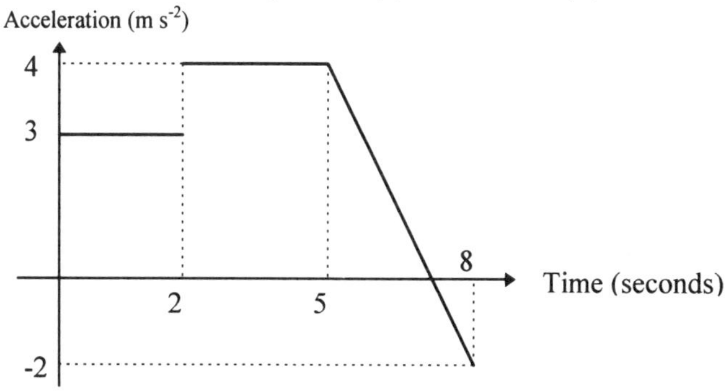

Work out where the line cuts the time axis.

(a)

Constant acceleration.
$v = u + at$ <u>or</u> initial velocity + area

(b)

Constant acceleration
$v = u + at$ <u>or</u> initial velocity + area

(c)

initial velocity + area above t axis - area below t axis.

Vertical Motion:

This is motion in a straight line (up and down) with constant acceleration due to gravity. Use $g = 9.8\ ms^{-1}$ unless told otherwise. Use the equations of motion as before but this time you know the acceleration.

Note: (i) If an object is thrown up, at the top of its flight its velocity is zero.
The velocity of a particle *as it hits the ground* is NOT zero. (think of the velocity just before it hits the ground)
Choose a positive direction (Usually up so, $a = -9.8$)

1. Complete: You throw a ball vertically upwards with an initial velocity of $15\ m\ s^{-1}$.

(i) Find the time taken for the ball to reach the top of its trajectory
(ii) Find the time taken for the ball to return to your hand
(iii) If you do not attempt to catch the ball and it hits the ground 1.2 m below your hand, how long has it been in the air?
(iv) How fast does it hit the ground?

Model the ball as a particle.

Units will be ms^{-1}, m, s and ms^{-2}

Answers to 3 sig. fig.

(i) Up positive: $u = 15 \quad v = 0 \quad a = -9.8 \quad t = ?$
Will use $v = u + at$

$0 = 15 + (-9.8)t \Rightarrow t =$ secs.

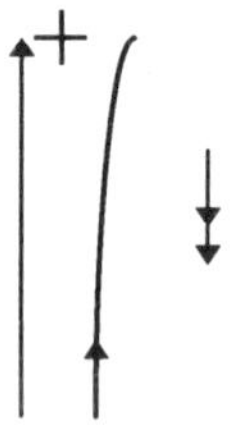

The initial velocity is given as $15\ ms^{-1}$ and the ball was thrown up, so up must be the positive direction.
List the values you know and the one you want to find. This should help you decide which equation of motion you need to use.

(ii) Up positive: $u = 15 \quad s = 0 \quad a = -9.8 \quad t = ?$
Will use $s = ut + \frac{1}{2}at^2$

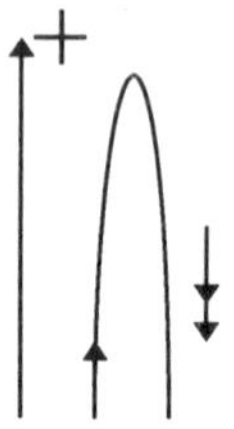

$0 =$

$0 = t\,(15 - 4.9t)$

$\Rightarrow t = 0$ (when ball is thrown) or $t = \quad = \quad$ secs

So ball returns to hand level after secs.

Displacement at the hand is 0 : this will happen twice - first when the ball is thrown and again when the ball falls back to its initial position
Factorise - t is a common factor

(iii) Up positive: $u = 15 \quad s = -1.2 \quad a = -9.8 \quad t = ?$
Will use $s = ut + \frac{1}{2}at^2$

Position is 1.2 m below the initial position - therefore if up is +ve then $s = -1.2$

$-1.2 = 15t - 4.9t^2 \Rightarrow 4.9t^2 - 15t - 1.2 = 0$

$t =$

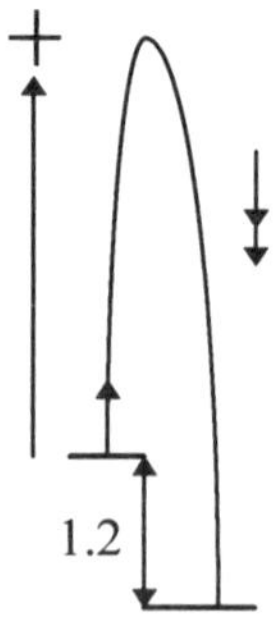

Use the quadratic formula
Use the positive root

(iv) Up positive: $u = 15 \quad s = -1.2 \quad a = -9.8$
Will use $v^2 = u^2 + 2as$
$v^2 = 225 + 2\,(-9.8)(-1.2) =$

Square root

$v =$

2. Complete: A ball is dropped from the top of a cliff. 1 second later a stone is thrown vertically down from the top of the cliff at 15 ms^{-1}. The ball and the stone hit the sea at the same time. Find the height of the cliff. Ball: $u = 0$, $a = 9.8$, $t = T$ $s = ut + \frac{1}{2}at^2$ Stone: $u = 15$, $a = 9.8$, $t = T - 1$ $s = ut + \frac{1}{2}at^2$ $s = 15(T-1) + 4.9(T-1)^2$ $s =$	*Take down as positive* *If the ball is in the air for T seconds, the stone is in the air for 1 second less i.e. T-1 seconds* *Careful* $(T-1)^2 = (T-1)(T-1)$ *Distance the same, so equate to find the time when they are together.* *Use this time to find the height (the displacement) at this time.* *Answer to 3 sig. fig.*
3. A book falls from a shelf 2 m above the floor. At what speed does it hit the floor?	*Take down as positive* $u=0$ $s=2$ $a=9.8$ *Use* $v^2 = u^2 + 2as$
4. A stone is dropped from a cliff 70 m above the sea. How long is it in the air before it hits the sea?	*Take down as positive* $u = 0$ $s = 70$ $a = 9.8$ $s = ut + \frac{1}{2}at^2$ *Answer to 3 sig. fig.*
5. A brick is dislodged from a hod on some scaffolding 25 m above the ground. Its initial velocity is 0.5 m s^{-1} downwards. Find the velocity with which it hits the ground and the time it takes to reach the ground.	*Down positive* $u = 0.5$ $s = 25$ $a = 9.8$ *Use* $v^2 = u^2 + 2as$ *for final velocity* *Then* $v = u + at$ *or* $s = ut + \frac{1}{2}at^2$ *for time to fall to ground.* *Answer to 3 sig. fig.*
6. A stone is thrown upwards with a speed of 20 m s^{-1}. Find the length of time the stone is above a height 10 m, if the stone is thrown from ground level the stone is thrown from 2 m above ground level	*Use* $s = ut + \frac{1}{2}at^2$ *to find the 2 times when the stone is 10 m above the ground* *Take up as positive* *For (ii) If the stone is thrown from zero displacement, you need the times when the displacement from the initial position is 8 m.* *Find the 2 times when the stone has a displacement of 8 m* *Answer to 3 sig. fig.*

FORCES AND NEWTON'S LAWS

The unit of force:
The unit of force is the Newton.

1 Newton = 1 kg m s^{-2} (1 kilogram metre per second squared)

This is the force required to accelerate 1kg at 1metre per second per second.

Force Diagrams:
With questions on forces always draw a diagram showing the forces on the body you are considering.
If the diagram is correct then you will be able to solve the problem correctly.
You must consider:
Weight - this is a force and equal to mg where m is the mass of the object. Use g = 9.8 m s^{-2} unless you are told otherwise.

Contact forces - put these in normal (perpendicular) to the surface of contact.

Pushing (thrust) and pulling forces - do not forget these if you are told about them.

Driving force - this is the force that drives the car/boat/bicycle/motorised toy etc.

Resistive forces - these forces oppose the motion of the object (mainly friction). If you have an object on a smooth surface then there is no friction.

Draw clear diagrams to show the forces acting on the following objects.
Consider the underlined objects only. All objects are considered as particles or rigid and uniform bodies.

A parcel lying on a table

A conker hanging at the end of a piece of string

Two books, one on top of the other, lying on a table.

(a) the book on the top

(a) the book on the bottom

A plank lying horizontally, resting on a brick at each end

A ball in mid air, having been thrown from one person to another (ignore air resistance)

A roof tile with one end on horizontal ground and the other leaning on a vertical wall

A pencil being held vertical with its base on a horizontal table and its point held by a downward pressure from your finger

A horse pulling a cart along a horizontal road

A car pulling a trailer

A car pulling a trailer

Resolving forces:

A force is a vector: it has magnitude (size) and direction. When working with a force we can write it in different ways

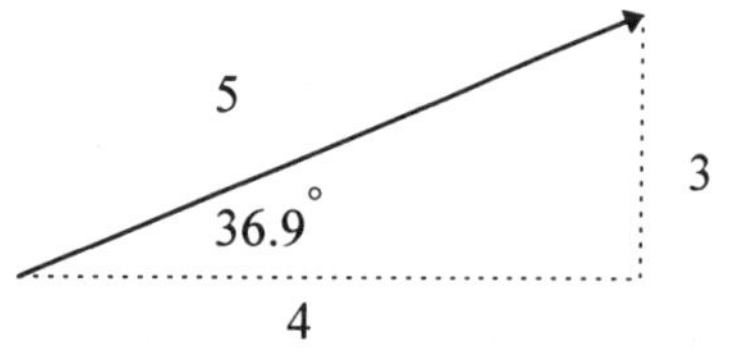

Magnitude / direction form

5 Newtons at 36.9° to horizontal

Component form

$(4\mathbf{i} + 3\mathbf{j})$ Newtons or $\begin{bmatrix} 4 \\ 3 \end{bmatrix}$ Newtons

To put a force in component form, the components do not need to be horizontal and vertical (but easier to deal with if they are perpendicular).

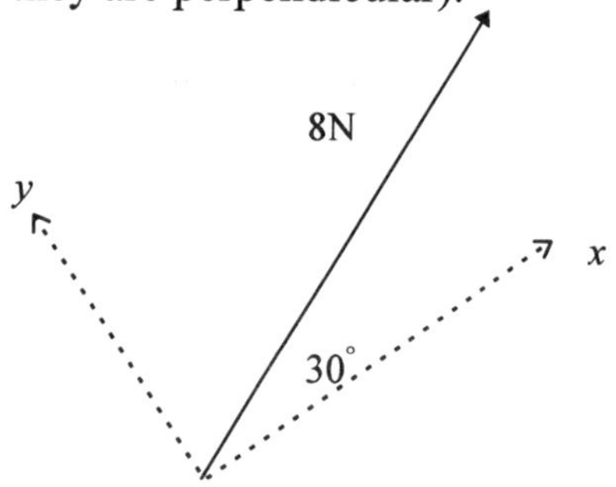

x component $8\cos30^\circ$
y component $8\sin30^\circ$

OR $(8\cos30^\circ\mathbf{i} + 8\sin30^\circ\mathbf{j})$ Newtons

OR $\begin{bmatrix} 8\cos30^\circ \\ 8\sin30^\circ \end{bmatrix}$ Newtons

This means that the original force is equivalent to (has the same effect as) two perpendicular forces. i.e.

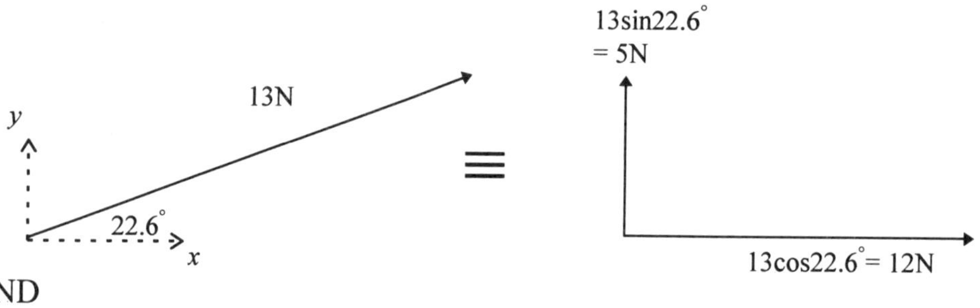

AND

Complete:

Resolve this forces in the given directions

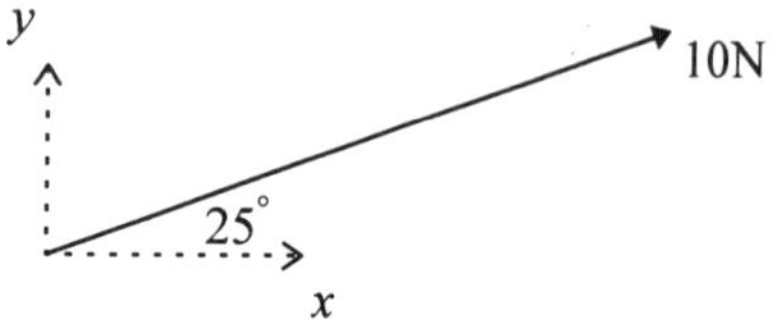

Horizontal component =

Vertical component =

Or written as a vector

$\begin{bmatrix} \quad \\ \quad \end{bmatrix}$

You could be given forces already in component form. e.g. $3\mathbf{i} - 4\mathbf{j}$

1. Resolve these forces in the given directions (Give answers to 3 s.f.)

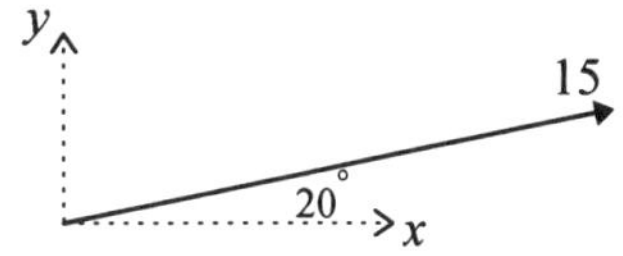

You can think '15cos the angle you are turning through.'.
(15cos 80° is the same as 15sin20°)
Write as separate components or as a column vector.

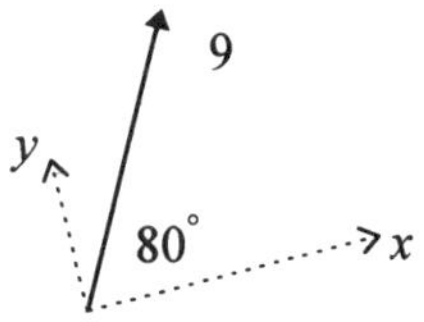

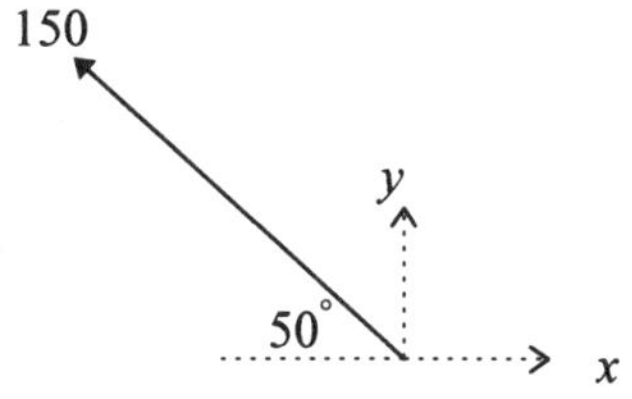

x component will be negative.

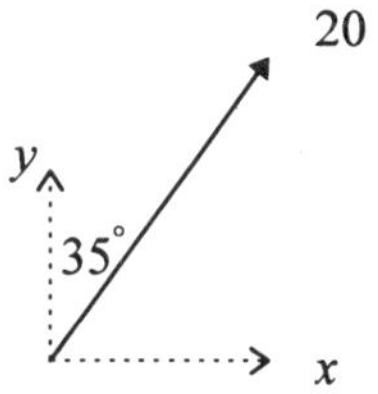

Think '20cos the angle you are turning through.'.

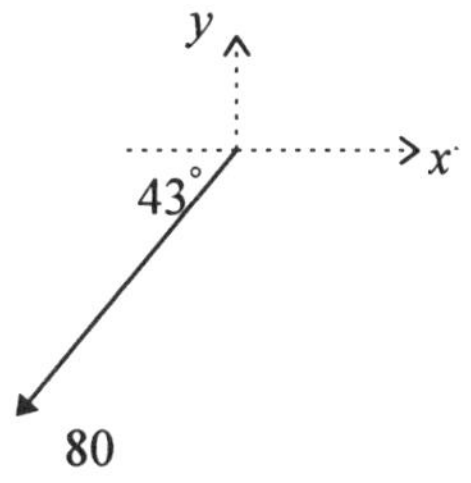

Careful.
Both components will be negative.

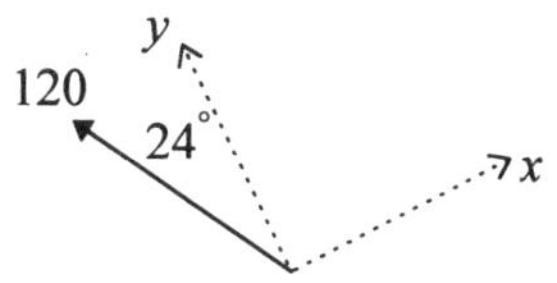

One component negative.

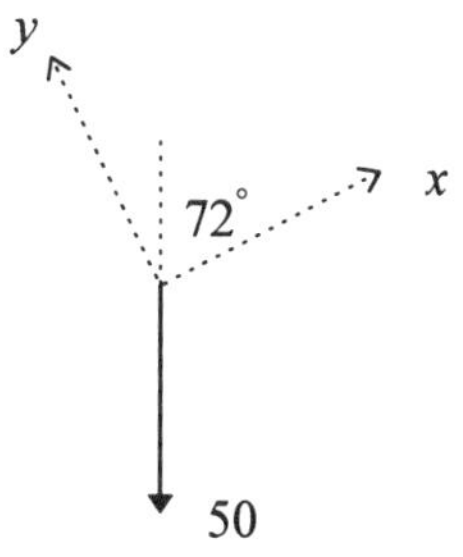

Both components will be negative.

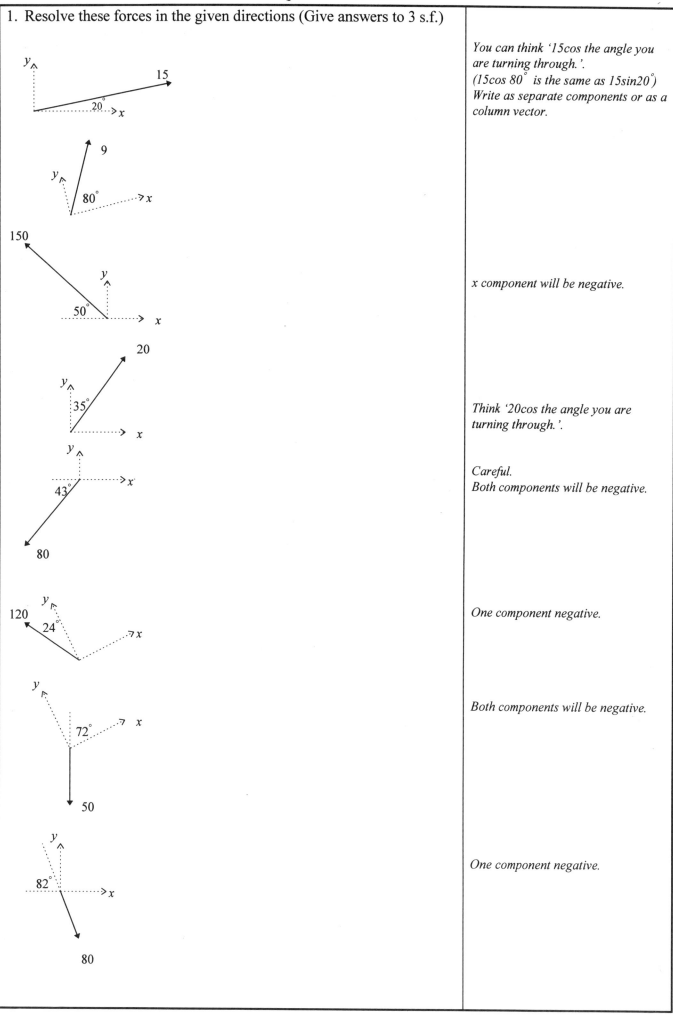

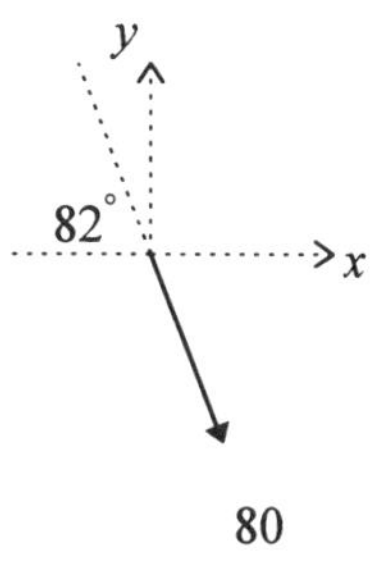

One component negative.

Magnitude and direction form from component form:

When solving problems you may end up with you answer in component form and need to work out the magnitude and or direction of the resultant vector.

e.g. (2.3**i** - 5.7**j**) N

Always sketch the resultant. Here 2.3 in the **i** direction followed by -5.7 in the **j** direction.
(This is to ensure that you get the direction correct)

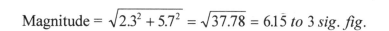

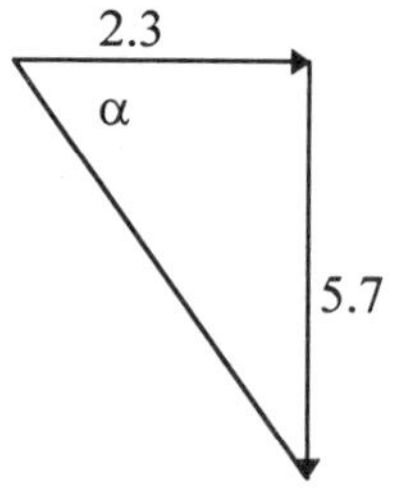

No need to write -5.7. The direction shows that it is -ve.

Magnitude = $\sqrt{2.3^2 + 5.7^2} = \sqrt{37.78} = 6.15$ *to 3 sig. fig.*

$\tan\alpha = \frac{5.7}{2.3} = 2.47826$

$\alpha = 68.0^\circ$

The direction is 68.0° clockwise from the **i** direction.

Use tan to find the angle. Be clear in your answer. -68° is acceptable (clockwise is -ve and anti-clockwise is +ve).

1.Find the magnitude and direction of the following forces:

First sketch the force vector.
*5 in the **i** direction followed by 8 in the **j** direction.*

Then find the magnitude and direction.
Be clear in giving your direction.

Answers to 3 sig. fig.

(5**i** + 8**j**)N

(9**i** - 2**j**)N

(-7**i** + 5**j**)N

(-3**i** - 8**j**)N

A resultant force is -77 N in the **i** direction and 42 N in the **j** direction.

The resultant force:

For forces acting in one dimension (along a straight line) just add and/or subtract the forces to give the resultant force. Make one direction positive. State the direction of the resultant.

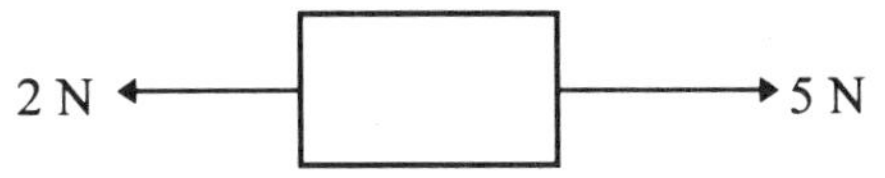

Resultant force = 5 - 2 = 3 N in the 5 N direction.

Resultant force = 6 + 8 - 12 = 2 N in the 6 N direction.

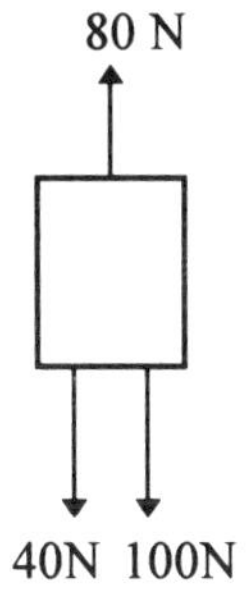

Resultant force = 100 + 40 - 80= 60 N downwards.

For forces given in component form again simply add and/or subtract.

The resultant of the forces (3**i** + 2**j**) N, (5**i** - 3**j**) N and (2**i** + 7**j**) N is:

3**i** + 2**j** + 5**i** - 3**j** + 2**i** + 7**j** = (10**i** + 6**j**) N   or as column vectors $\begin{bmatrix}3\\2\end{bmatrix}+\begin{bmatrix}5\\-3\end{bmatrix}+\begin{bmatrix}2\\7\end{bmatrix}=\begin{bmatrix}10\\6\end{bmatrix}$ N

This can easily be shown graphically

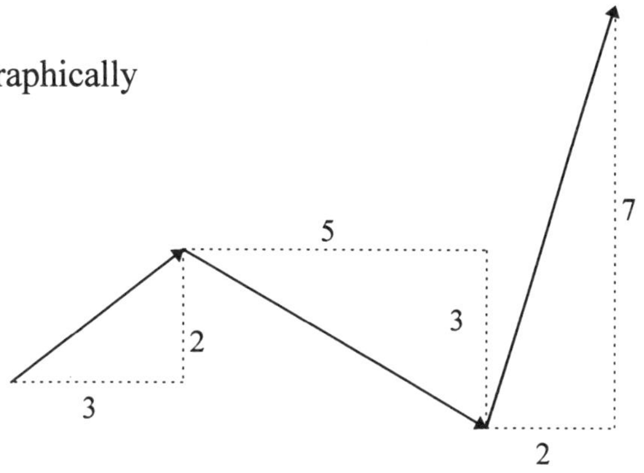

In **i** direction 3+5+2=10

In **j** direction 2-3+7=6

1. Find the resultant of the forces (-5**i** + 2**j**) N, (8**i** + 3**j**) N and (2**i** - 7**j**) N.	*Just add.*
2. Work out the resultant force in each of the following cases. 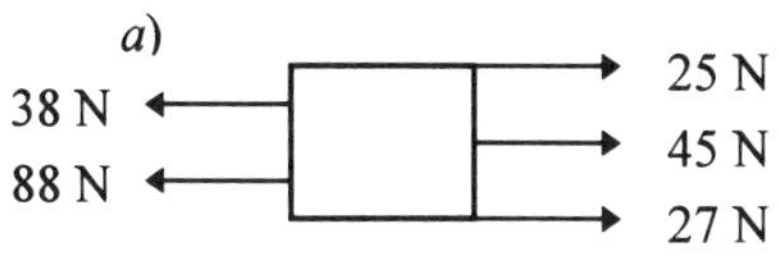 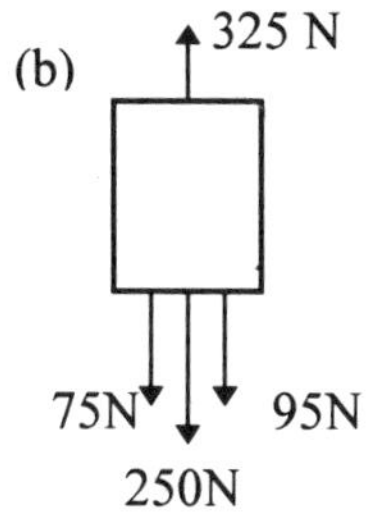	*For forces acting along a straight line. Make one direction positive and add.*

For forces acting in two dimensions but not given in component form you just need to work out the components in two perpendicular directions and then the problem becomes just one of adding the components as above.

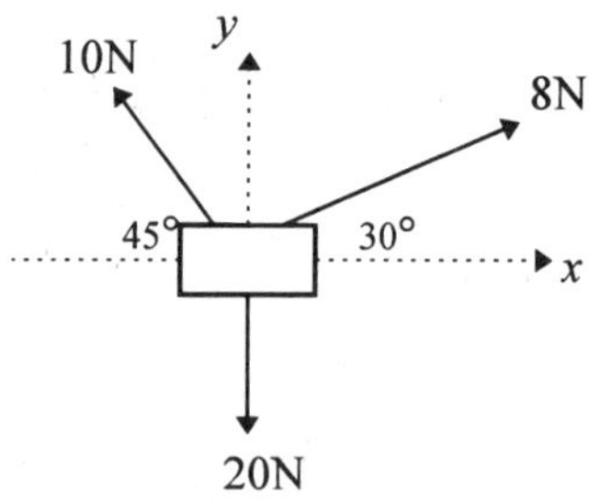

First label (x and y) the directions that you are going to use when resolving the forces.

Then sum the horizontal components of all the forces.

$8\cos30^\circ - 10\cos45^\circ + 0 = -0.14286$

and sum of the vertical components

$8\sin30^\circ + 10\sin45^\circ - 20 = -8.92893$

OR

Write down the vector for each force and sum them

$$\begin{bmatrix} 8\cos30^\circ \\ 8\sin30^\circ \end{bmatrix} + \begin{bmatrix} -10\cos45^\circ \\ 10\sin45^\circ \end{bmatrix} + \begin{bmatrix} 0 \\ -20 \end{bmatrix} = \begin{bmatrix} -0.14286 \\ -8.92893 \end{bmatrix}$$

1. Find the magnitude and direction of the resultant of the forces acting on this body.

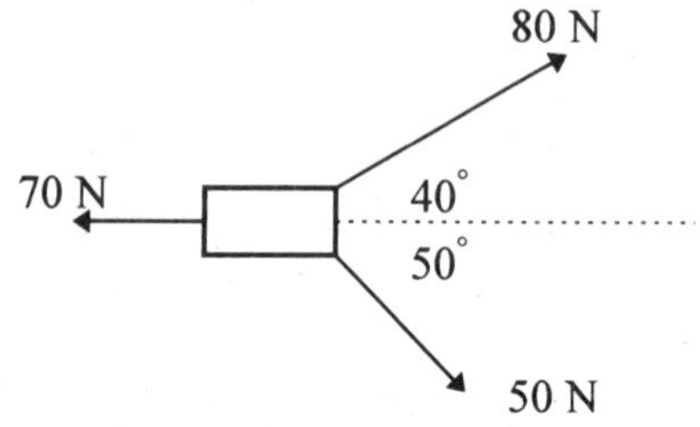

Resolving in the x direction

Resolving in the y direction

Sketch of the resultant.

Magnitude

Direction

First label (x and y) the directions that you are going to use when resolving the forces.

Sum of the components in the x and y directions.

OR write as the sum of vectors.

$$\begin{bmatrix} \quad \\ \quad \end{bmatrix} + \begin{bmatrix} \quad \\ \quad \end{bmatrix} + \begin{bmatrix} \quad \\ \quad \end{bmatrix} = \begin{bmatrix} \quad \\ \quad \end{bmatrix}$$

x direction followed by y direction.

Pythagoras

Tan of the angle
Make it clear - angle and rotation from which direction.

Answers to 3 sig. fig.

2. Find the resultant of the following forces in the form $a\mathbf{i} + b\mathbf{j}$.

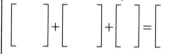

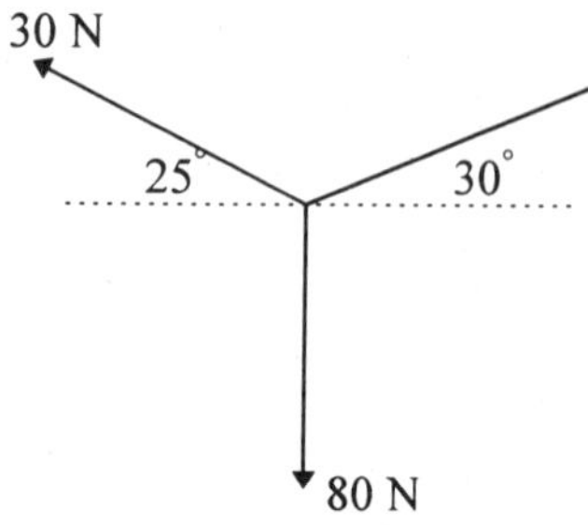

Resolving in the x direction

Resolving in the y direction

Answer:

First label (x and y) the directions that you are going to use when resolving the forces

Sum of the components in the x and y directions.
OR write as the sum of vectors.

Answer to 3 sig. fig.

3. Find the resultant of these forces

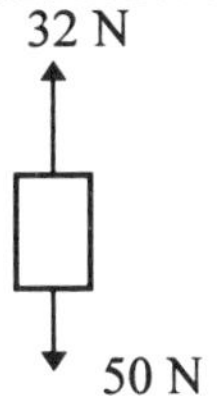

A simple question.

Don't forget the direction.

4. Find the resultant of these forces

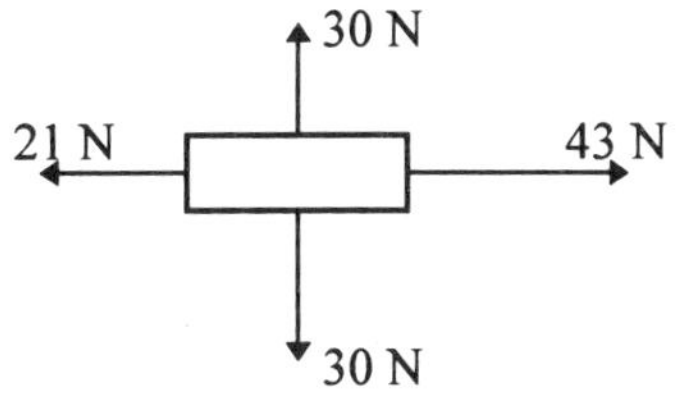

Resolving in the x direction

Resolving in the y direction

Answer:

A simple question. The forces are already in the x and y directions. Just sum in each of these directions.

Be clear in your answer.

5. Find the resultant of these forces

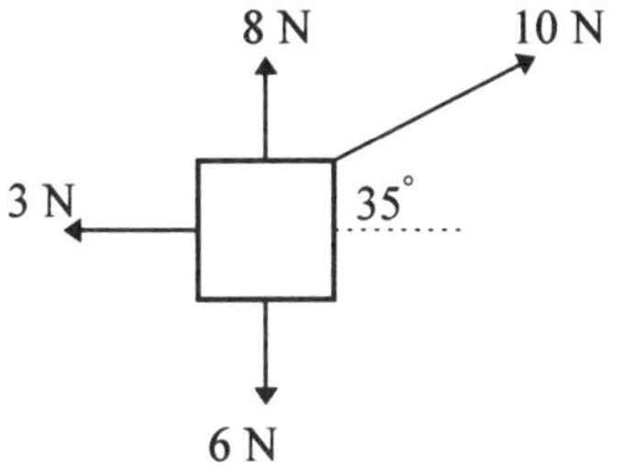

Choose and label the x and y directions.
Resolve.

Answers to 3 sig. fig.

6. Find the resultant of these forces

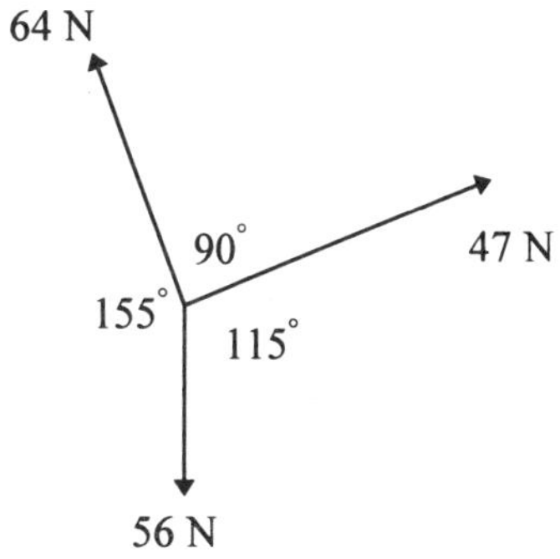

Notice the 47 N and the 64 N forces are at right angles. So choose these directions for the x and y directions.

Work out the angle of the 56 N force to these directions.

Sum the components in these directions.

Answers to 3 sig. fig.

7. A rock is being dragged up a hill as shown. Find the resultant force up the hill.

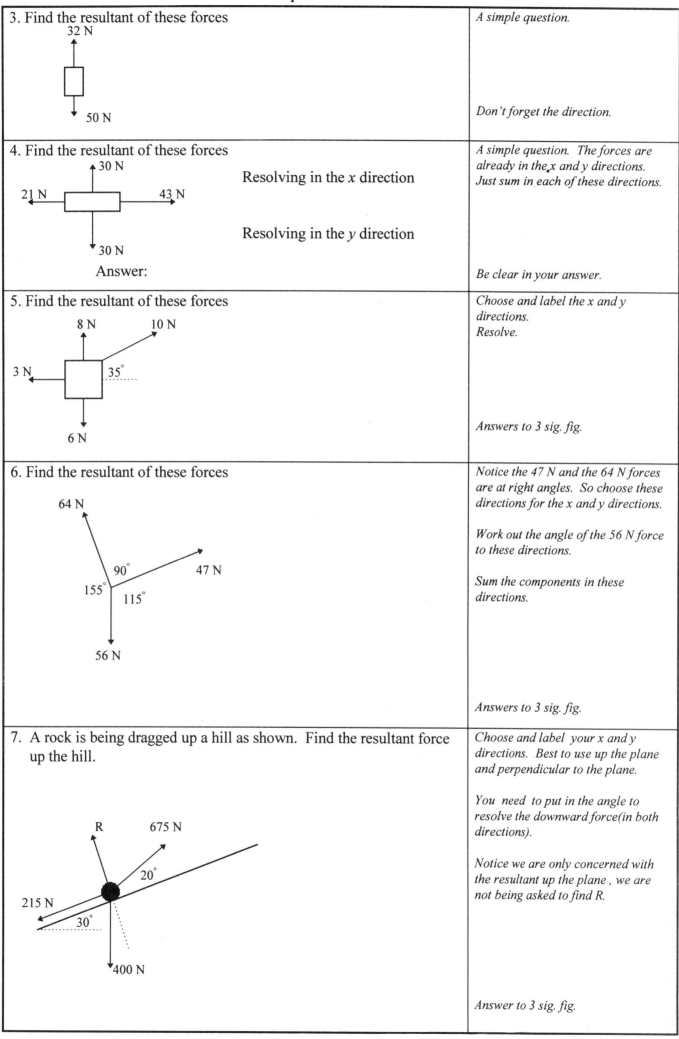

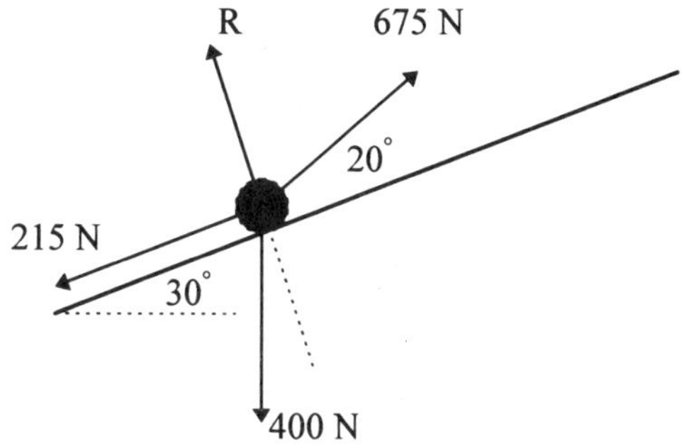

Choose and label your x and y directions. Best to use up the plane and perpendicular to the plane.

You need to put in the angle to resolve the downward force(in both directions).

Notice we are only concerned with the resultant up the plane , we are not being asked to find R.

Answer to 3 sig. fig.

Newton's First Law of Motion:

A body will continue in a state of rest or move at a constant velocity unless acted upon by an external force.
In other words a change in the velocity of a body is caused by a force.

Equilibrium:

<u>A body at rest:</u> - if forces act on the body and it does not move the forces must be "in equilibrium" - they balance out.
The resultant force IN ANY DIRECTION is zero.

<u>A moving body:</u> - If the body is moving at a <u>constant velocity</u> there is no resultant force acting on it. i.e. the forces must be "in equilibrium" - they balance out.
The resultant force IN ANY DIRECTION is zero.
(If a resultant force acts on it its velocity will change: the body may speed up, slow down or change direction.)

> If a set of forces are in equilibrium the RESULTANT FORCE (sum of the forces) must be zero. And so the sum of their components, in any direction, must also be zero.

Forces acting in one dimension

A body is at rest under the action of the forces shown in the diagram. Fill in the value of A. 	*The sum of the forces is zero (Make one direction +ve)* *OR* *the forces acting to the right equals the forces acting to the left.*
A body moves horizontally at a constant 4 ms^{-1} under the action of the forces shown. Find X 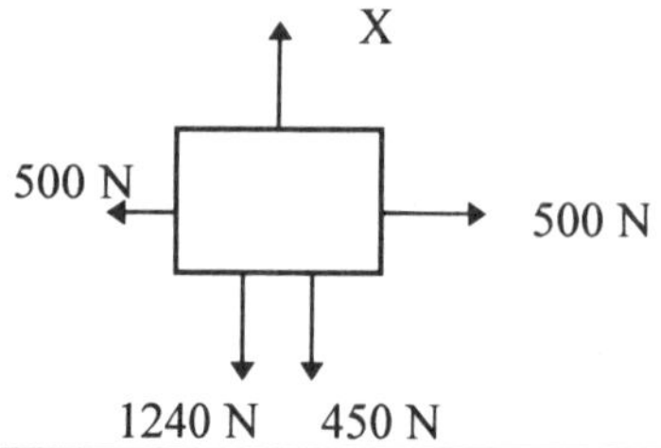 X =	*The sum of the forces is zero (make one direction +ve)* *OR* *the forces acting up equals the forces acting down.*

Forces acting in 2 dimensions.

A body moves horizontally at a constant 4 ms^{-1} under the action of the forces shown. Find X and Y 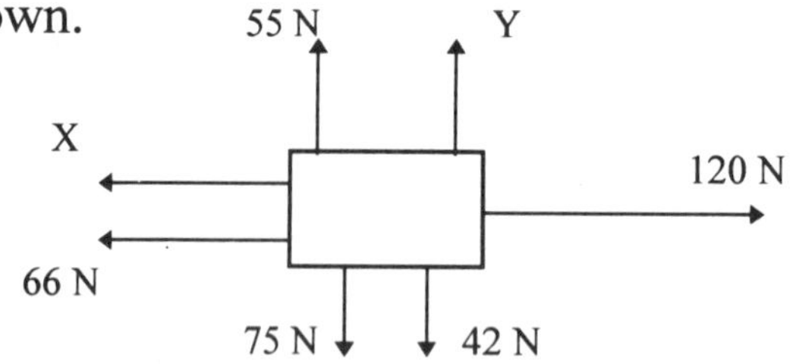 X = Y =	*No acceleration in any direction. Therefore the forces are in equilibrium.*

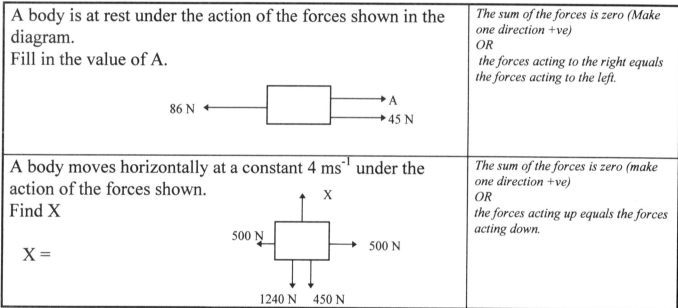

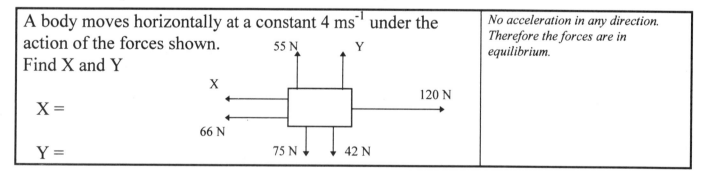

For forces given in component form.

Example: The forces (6**i** - 4**j**) N, (-2**i** + 3**j**) N and (x**i** + y**j**) N are in equilibrium. Find x and y. Parallel to x axis (in direction of **i**): $6 - 2 + x = 0$ $x = -4$ Parallel to y axis (in direction of **j**): $-4 + 3 + y = 0$ $y = +1$	*The forces are already split into components here. A diagram is not necessary.*

1. A body of mass 20 kg is being pulled along the floor with constant velocity by a horizontal force of magnitude 140 N. Find the normal reaction acting on the box and the resistance to motion.	*Draw a diagram.* *Constant velocity - therefore the forces are in equilibrium.* *Remember, weight = mg.*
2. A person of mass 45 kg is in a lift. The lift is going vertically upward with a constant velocity. What is the reaction force of the lift on the person?	*Draw a diagram. Put in the forces acting on the person.* *Constant velocity - therefore the forces are in equilibrium.*
3. Forces $2\mathbf{i} + 4\mathbf{j}$, $5\mathbf{i} - q\mathbf{j}$, $-2\mathbf{i} + 5\mathbf{j}$ and $p\mathbf{i} - 6\mathbf{j}$ newtons are in equilibrium. Find the value of p and q.	*In equilibrium. Therefore thsum of the forces equals zero.* *Total of* $\mathbf{i}$ *components must be zero.* *Total of* $\mathbf{j}$ *components must be zero.*
4. Each of these sets of forces are in equilibrium. Find the values of x and y in each case. (i) $(5\mathbf{i} + 3\mathbf{j})$ N, $(2\mathbf{i} - 7\mathbf{j})$ N, $(x\mathbf{i} + y\mathbf{j})$ N (ii) $(-3\mathbf{i} - 8\mathbf{j})$ N, $(x\mathbf{i} + 3\mathbf{j})$ N, $(-2\mathbf{i} + y\mathbf{j})$ N $(\mathbf{i} - \mathbf{j})$ N, $(3\mathbf{i} - y\mathbf{j})$ N, $(-5\mathbf{i} - y\mathbf{j})$ N, $(x\mathbf{i} + 5\mathbf{j})$ N	*Total of* $\mathbf{i}$ *components must be zero.* *Total of* $\mathbf{j}$ *components must be zero.*

For forces not given in component form.

It is important to draw a diagram showing all the forces acting on the body if you need to resolve the forces.

For any problem involving forces in equilibrium you need to look at the sum of the forces in two perpendicular directions. For equilibrium the sums in these directions must be zero.

These forces are in equilibrium

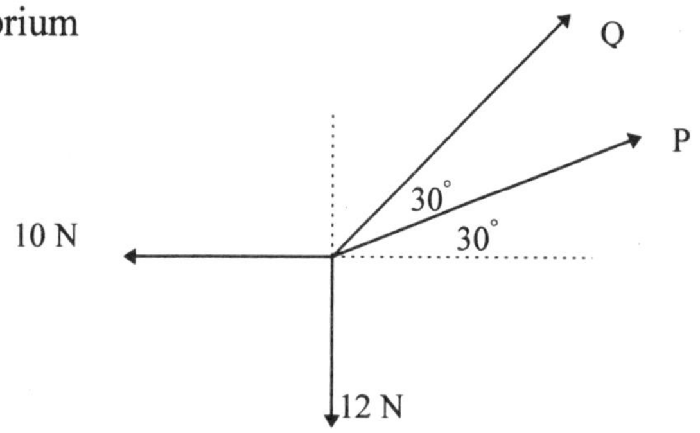

Complete:

First label (x and y) the directions that you are going to use when resolving the forces

Write down the x direction components and sum (add) them. In equilibrium, so will equal zero. (OR forces to the right = forces to the left,)

Write down the y direction components and sum (add) them. In equilibrium so will equal zero. (OR forces up = forces down)

We can then use these two equations to find P and Q.
Solve the simultaneous equations.

OR

First label (x and y) the directions that you are going to use when resolving the forces
Write down the vector for each force and sum them. In equilibrium so will equal zero.

$$\begin{bmatrix} \quad \\ \quad \end{bmatrix} + \begin{bmatrix} \quad \\ \quad \end{bmatrix} + \begin{bmatrix} \quad \\ \quad \end{bmatrix} + \begin{bmatrix} \quad \\ \quad \end{bmatrix} = \begin{bmatrix} 0 \\ 0 \end{bmatrix}$$

Now write down the two equations which you can use to help to solve problems.

$= 0$

$= 0$

Then follow working in left hand box from

←

Example: One end of a light inextensible string is attached to a horizontal ceiling and a particle of 4 kg is attached to the other end of the string. The particle is pulled sideways by a horizontal force of 20 N until the equilibrium position is reached.
Find the tension in the string, T and the angle which the string makes with the vertical (in the equilibrium position) θ.

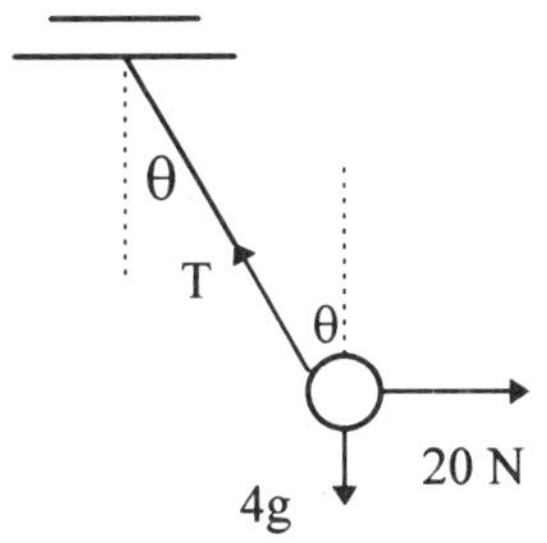

Resolve vertically: $T\cos\theta = 4g = 4 \times 9.8$
$= 39.2$ N (i)

Resolve horizontally: $T\sin\theta = 20$ N (ii)

(ii)/(i) $\tan\theta = \dfrac{20}{39.2}$ so $\theta = 27.0^\circ$ (3 s.f.)

in (i) $T\cos\theta = 39.2$ so $T = \dfrac{39.2}{\cos 27.0^\circ}$
$= 44.0$ N (3 s.f.)

A good diagram helps you see the directions of all the forces

The most sensible choice for directions of resolving is horizontal / vertical

OR sum of forces is zero.

$$\begin{bmatrix} -T\sin\theta \\ T\cos\theta \end{bmatrix} + \begin{bmatrix} 20 \\ 0 \end{bmatrix} + \begin{bmatrix} 0 \\ -4g \end{bmatrix} = \begin{bmatrix} 0 \\ 0 \end{bmatrix}$$

So -

$T\sin\theta + 20 = 0$

and

$T\cos\theta - 4g = 0$

Then follow working on the left.

Resolving forces in the horizontal and vertical directions is usually the most convenient - but not always. When a body is in equilibrium on an inclined plane it is usually sensible to resolve forces parallel to the surface of the plane and perpendicular to the surface of the plane.

1. Complete: A brick of mass 6 kg rests on a sloping roof inclined at 30° to the horizontal. A frictional force F acting up the plane holds the brick at rest on the roof. Find F

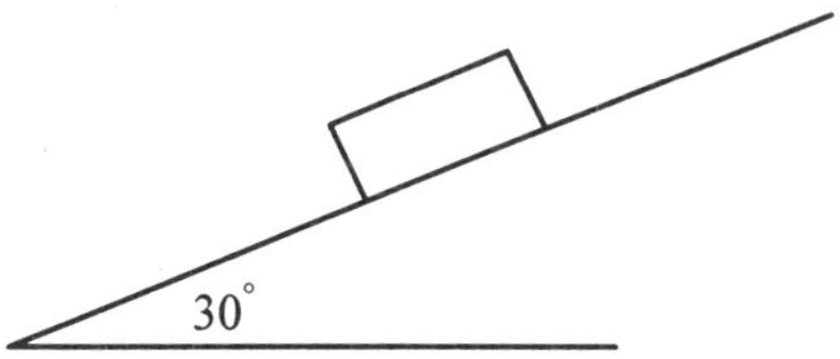

Resolve parallel to the plane:
$F = 6g\sin 30^\circ$
$= 6 \times 9.8 \times 0.5$
$= 29.4$ N

The brick is considered as a particle. The roof is considered as a plane.

Put in the forces acting on the brick. i.e. F, mg, and R

Put in the x and y directions. Here parallel and perpendicular to the plane.

Put in the angle that the mg makes with the perpendicular to the plane.

The reaction force R is perpendicular to the plane and therefore has no component in the direction parallel to the plane.

2. A body of mass 8 kg is being held by two strings as shown. the angle between the strings is 90°. Find the tension in each string.

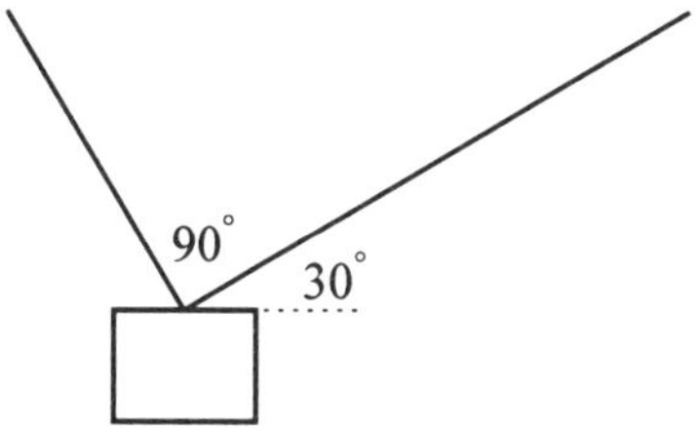

Put in the tensions T_1 and T_2 and the weight. (weight is mg = 8g)

The strings are at right angles, so use these directions as your x and y directions.

Work out the angle that the 8g force makes with the y direction.

Resolve in these directions.

Answer to 3 sig. fig.

3. A light inelastic string of length 20 cm has one end attached to a fixed point A and at the other end a particle of mass 2 kg is fixed. Consider 2 cases:

Answers to 3 sig. fig.

A horizontal force P is applied to the particle and keeps it in equilibrium 12 cm from the vertical through A.
Find P and the tension in the string, T.

The triangle made by the string, the vertical through A and the horizontal through the particle is a right angled triangle with sides in the ratio 3:4:5
Find θ

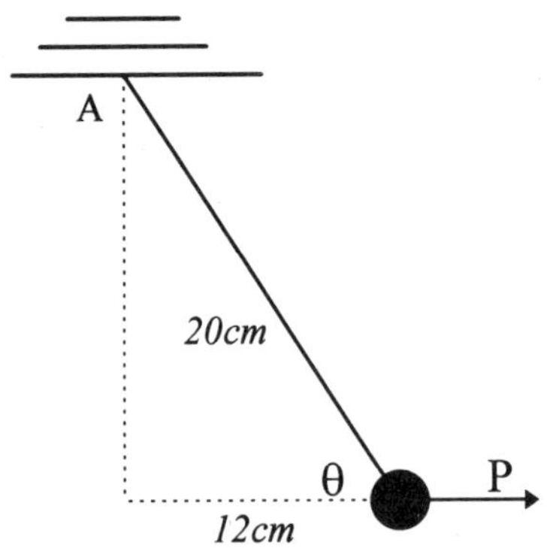

Put in the forces. T and the weight.

Resolve vertically to find t.
Resolve horizontally to find P.

A force Q perpendicular to the string is applied to the particle and keeps it in equilibrium 10 cm from the vertical through A.
Find Q and the tension in the string, T_Q

Draw a diagram.
Add in all the forces
Find the new θ
Resolve perpendicular to the string to find Q.
Resolve parallel to the string to find T_Q

4. A toy sledge of mass 4 kg on a snowy hill is modelled as a particle on a smooth plane inclined at 25° to the horizontal. A child is holding the sledge steady by putting his foot at the front of the sledge. This is modelled as a force A newtons acting parallel to and up a line of greatest slope. Find A and the normal force between the sledge and the hill.	*Draw a diagram to show the forces acting on the sledge.* *Answers to 3 sig. fig.*
5. The same sledge as in the previous question is now moved to a steeper slope inclined at 35° to the horizontal. This time it is held at rest by a horizontal force, B newtons. Find B and the reaction force between the sledge and the plane.	*Draw a diagram to show the forces acting on the sledge.* *Answers to 3 sig. fig.*
6. You are holding a model car on a ramp inclined at angle α to the horizontal, where $\tan\alpha = \frac{5}{12}$. The car has mass 2 kg and you have to apply a force of 5 N parallel to the plane to just prevent the car from moving down the plane. If you model the car as a particle and the ramp as a rough plane, find the reaction force between the car and the plane and the friction between the car and the plane. 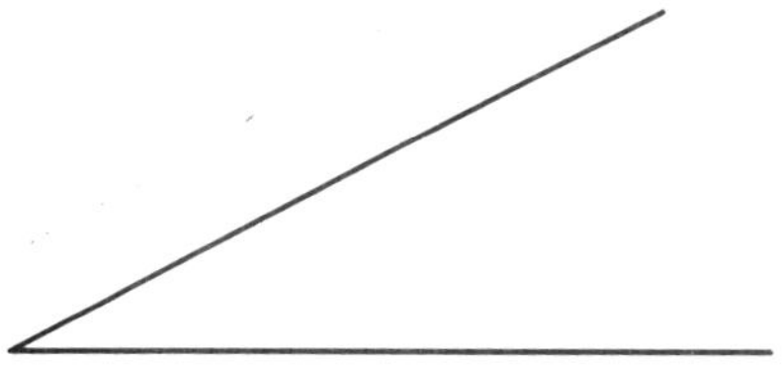	*Draw a right angled triangle containing* α to find $\cos\alpha$ and $\sin\alpha$. *"just prevents the car from moving down the plane" means motion would be DOWN the plane, so friction will act UP the plane.* *Resolve perpendicular to the plane.* *Resolve parallel to the plane.* *Answer to 3 sig. fig.*

7. A boat of mass 200 kg is being pulled down a sloping beach at a constant velocity towards the sea. The beach slopes at 15° to the horizontal. The friction between the boat and the beach is 700 N. The boat is pulled by a single rope running parallel to the greatest slope of the beach. Find the pulling force, P newtons, and the normal reaction force from the beach

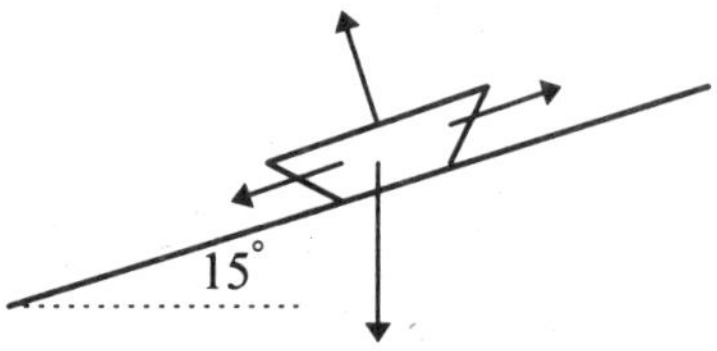

Complete the diagram

Answers to 3 sig. fig.

8. Two girls are pulling a sledge over a horizontal snowy field. One girl pulls with a force of 70 N, the other with a force of T N. There is no resultant force perpendicular to the line of symmetry of the sledge. Find T.

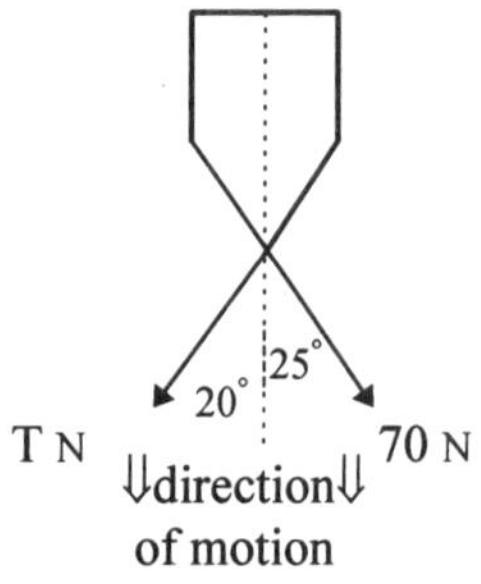

If the sledge is moving at a constant velocity along the direction of the line of symmetry of the sledge, find the drag force on the sledge.

Answers to 3 sig. fig.

Resolve perpendicular to the line of symmetry of the sledge. No resultant force, so equilibrium.

Not all the forces on the sledge have been put on the diagram. Add in the drag force.
Find it by resolving parallel to the direction of motion. Constant velocity means equilibrium again.

9. Two tugs are pulling a ship into harbour at a constant speed.

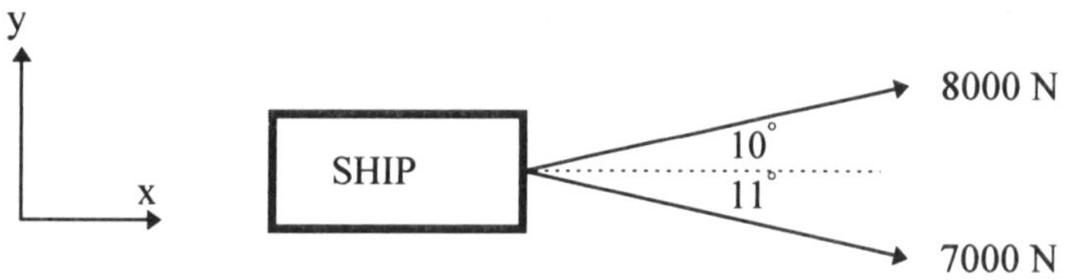

Resolve the forces in the x and y directions

X =

Y =

What is the overall force in the positive x direction?

What is the overall force in the positive y direction?

If the ship moves in the positive x direction, what can you say about resistance to motion parallel to the y axis?

resistance to motion parallel to the x axis?

Constant speed indicates equilibrium: no resultant force, therefore no acceleration.

Not all the forces acting on the ship are on this diagram. Complete the diagram.

Answers to 3 sig. fig.

Newton's Second Law of Motion:

Simply **Resultant force, F = m *a***

i.e. Resultant force in any direction equals mass times acceleration in that direction.

In problems you may have to find the acceleration from a given force system so that you can use the constant acceleration formulae to work out a displacement, velocity or time. In others you may have to use the constant acceleration formulae to work out an acceleration and hence a force by using F = m*a*.

Forces acting in one direction

1. Find the acceleration experienced by a particle of mass 6 kg when acted on by a resultant force of 25 Newtons.	*Just use F=ma.*
2. What resultant force will produce an acceleration of 0.5 m s^{-2} in a car of mass 850 kg?	*Just use F=ma*
3. Assuming acceleration to be constant find the force that increases the velocity of a car of mass 750 kg from 2 m s^{-1} to 8 m s^{-1} in 3 seconds.	*First find the acceleration using one of the constant acceleration formulae.* *Then use F=ma.*
4. A vehicle of mass 550 kg moving at 12 m s^{-1} is brought to rest in 5m. Find the braking force.	*First find the acceleration using one of the constant acceleration formulae.* *Then use F=ma.*
5. A trailer of mass 400 kg is being pulled along a straight road with a force of 550 N. If the resistance to motion is 375 N , find the acceleration of the trailer. 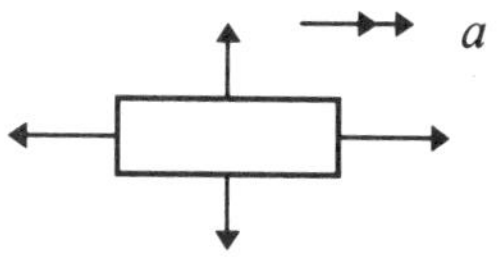 	*A sketch may help.* *No resultant force vertically. So no need to consider these forces.* *Remember <u>resultant</u> force = ma.*
6. The tension in a rope being used to raise an object is 61.5 N. If the acceleration of the object is 2.5 m s^{-2} find the mass of the object.	*Sketch.* *<u>Resultant</u> force = ma.* *Terms with m on one side of the equation so you can take m out as a factor.*
7. A bullet of mass 25 grams is fired horizontally into a fixed wooden block which offers a constant resistance of 1000 N to the motion of the bullet. Find the deceleration of the bullet.	*Newtons are kg m s^{-2} so the mass must be in kg.* *Then use F = ma.*

Forces acting in two perpendicular directions

1. A particle of mass 3 kg is acted on by a force of $(3\mathbf{i} + 7\mathbf{j})$ N. Find the resulting acceleration.

Use $\mathbf{F} = m\mathbf{a}$. $\mathbf{F}$ and $\mathbf{a}$ are vectors, so just put the vector for $\mathbf{F}$ in this equation and the acceleration will be a vector in this form.

2. Find the magnitude and direction of the resultant force required to give a particle of mass 5 kg an acceleration of $(5\mathbf{i} - 2\mathbf{j})$ m s^{-2}.

Use $\mathbf{F} = m\mathbf{a}$ to find $\mathbf{F}$ as a vector.
First sketch the force vector.
Then find the magnitude and direction.

Answer to 3 sig. fig.

3. Forces act on a body as shown giving rise to the acceleration also shown. Find the vertical force R and the horizontal force P.

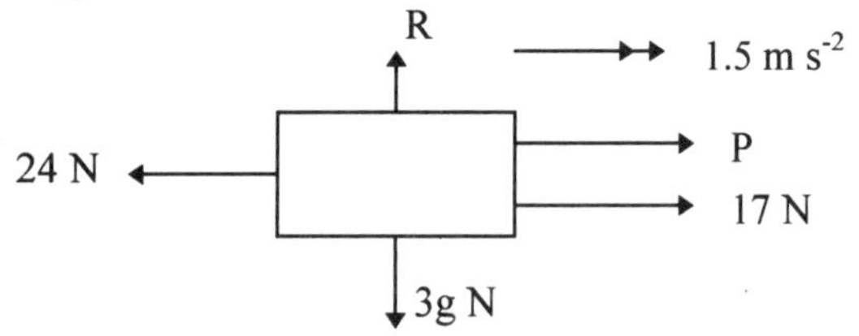

4. A particle of mass 8 kg is being pulled along a rough horizontal surface by a string parallel to the surface. If the tension in the string is 74 N and the resistance to motion is 60 N find the acceleration of the particle.

Sketch.
Put in the forces. mg, N (the normal reaction), T (tension), R (resistive forces)

Acceleration is a vector so state the direction.

Forces acting in two dimensions

1. A particle of mass 8 kg is being pulled along the ground by a string inclined at 30° to the horizontal. If the acceleration of the particle is 3 m s^{-2} and the resistance to motion is 40 N, find the tension in the string.

Resultant force = ma

=

=

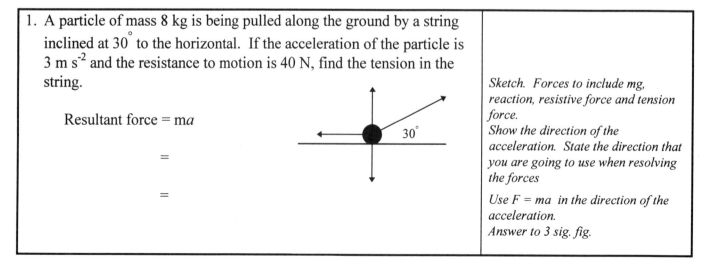

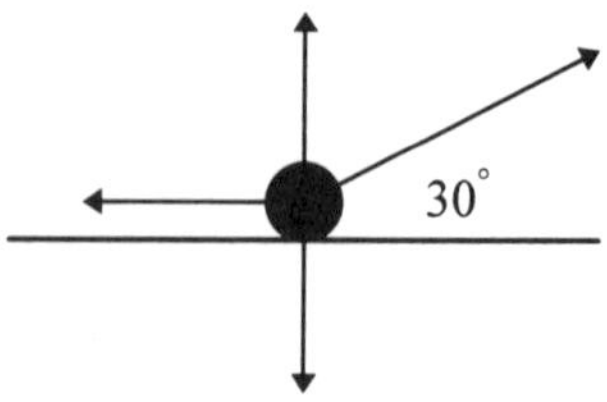

Sketch. Forces to include mg, reaction, resistive force and tension force.
Show the direction of the acceleration. State the direction that you are going to use when resolving the forces

Use F = ma in the direction of the acceleration.
Answer to 3 sig. fig.

2. A tile of mass 5 kg slides down a roof which is inclined at 40° to the horizontal. Assuming the roof is smooth find the acceleration of the tile.	*Sketch.* *Forces are weight and normal reaction.* *Show the direction of the acceleration. State the direction that you are going to use when resolving the forces.* *Answer to 3 sig. fig.*
3. A particle of mass 4 kg is being pulled up a smooth slope of 35° to the horizontal by a string inclined at 20° to the slope. If the acceleration of the particle is 0.5 m s^{-2} find the tension in the string.	*Sketch. Forces to include mg, reaction and tension force.* *Show the direction of the acceleration. State the direction that you are going to use when resolving the forces* *Use F = ma in the direction of the acceleration.* *Answer to 3 sig. fig.*
4. A ring of mass 4 kg slides down a wire inclined at 30° to the horizontal. The resistive forces to the motion of the ring total 3 N parallel to the wire. Find the acceleration of the ring	*Sketch. Forces to include mg and reaction normal to the wire.* *Show the direction of the acceleration. State the direction that you are going to use when resolving the forces* *Use F = ma in the direction of the acceleration.*
5. A boy of mass 20 kg slides down a straight slide inclined at 40° to the horizontal, against a resistance of 30 N. (a) Find his acceleration. (b) If the slide is 2 m long and the boy starts from rest, what is his speed when he reaches the end of the slide?	*Answers to 3 sig. fig.* *(a) Use F = ma* *(b) Use one of the constant acceleration formulae.*

Newton's Third Law of Motion:

If a body A exerts a force on a body B then B exerts an equal and opposite force on A

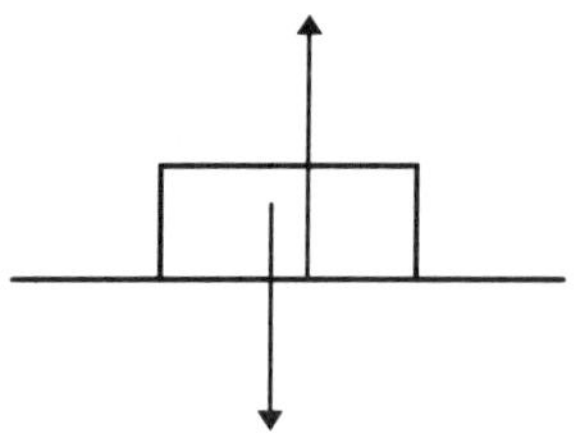

Body at rest. Body presses down on the ground and the ground pushes up on the body (reaction).

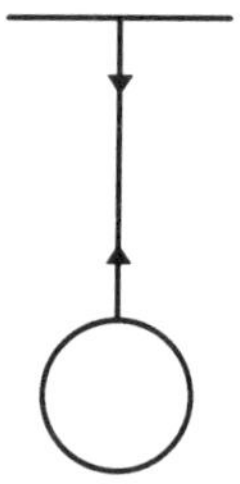

Body pulling down on the ceiling via string (tension) and the ceiling holding up the body via string (tension).

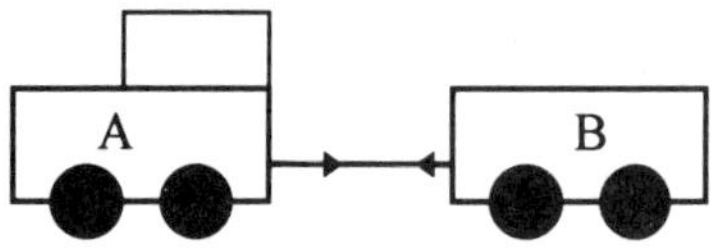

Body A pulling on B via string (tension) and Body B pulling on A via string (tension).

Connected bodies:

Remember: for a smooth pulley the tension in the string on either side of the pulley is the same.

1. Two sacks of grain of masses 35 kg and 25 kg are connected by a light inextensible string passing over a smooth pulley. If the sacks are released from rest, find the acceleration of the sacks and the tension in the string. For the 35kg sack: For the 25kg sack: 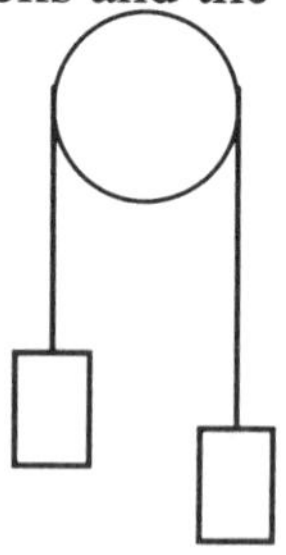	*Remember weight is mg* *Put in the forces acting on each sack. mg and the tension.* *Show the acceleration.* *Use resultant force = ma for each sack.* *Solve the simultaneous equations to find the acceleration and the tension.* *Answers to 3 sig. fig.*
2. Two objects A and B of mass 3 kg and 2 kg respectively, are connected by a string. Object A is held on a smooth horizontal surface and the string passes over a pulley so that object B hangs freely. Object A is released, find the acceleration of B and the tension in the string. List any other assumptions being made. 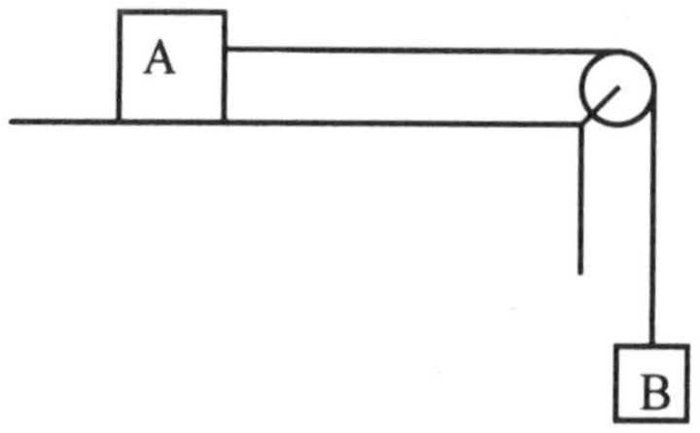	*Put in the forces acting on each object. mg, tension and normal reaction for A, mg and the tension for B* *Show the acceleration of A and B on the diagram. (The same for both . B down and A along)* *Use resultant force = ma for each object.* *The weight (mg) and normal reaction of A are not used in these calculations.*

3. Particle A of mass 5 kg is held on a rough surface. It is connected by a light inextensible string over a smooth pulley to a particle B of mass 11 kg which is hanging freely as shown. When A is released the particles begin to move. If the resistance to motion experienced by A is 30 Newtons, find
   (a) the acceleration of B
   (b) the tension in the string
   (c) the distance moved by A in 1 second assuming it does not hit the pulley.

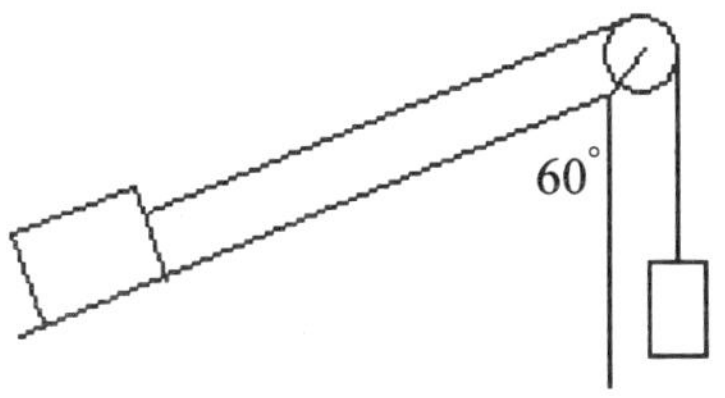

Put in the forces acting on each particle. mg, tension, normal reaction and resistance for A, mg and the tension for B

Show the acceleration of A and B on the diagram. (The same for both).

Use resultant force = ma for each object.

For A:
the direction of the acceleration is up the plane, so use this as your x direction and you will need to resolve the weight into its components. (To do this you will also need to work out the angle/s that the weight makes with the x and/or y directions.

Answers to 3 sig. fig.

4. Two blocks attached together by string are being pulled along a horizontal straight line by a force of 200 N as shown below. If block A has mass 7 kg and the resistance to its motion is 60 N and block B has mass 5 kg and the resistance to its motion is 40 N find the acceleration of the two blocks and the tension in the string.

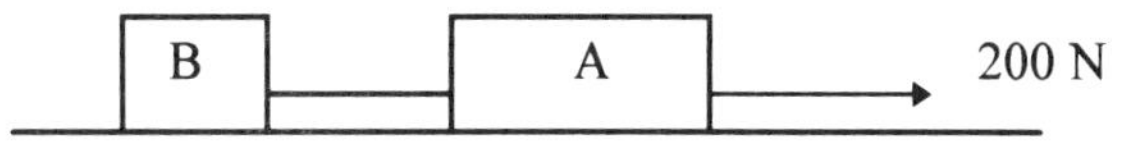

Put in the forces acting on each block.

Tension in string pulling back on A and pulling forward on B.

Show the acceleration of A and B on the diagram. (The same for both).

Use resultant force = ma for the whole system.

Then use resultant force = ma for either A or B to find the tension in the string.

Answers to 3 sig. fig.

5.

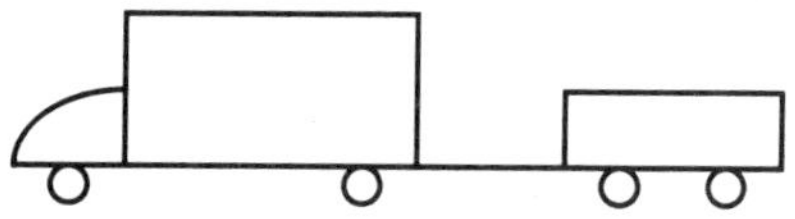

A car of mass m_1 kg is pulling a trailer of mass m_2 kg along a horizontal track. The engine of the car produces a forward driving force F newtons and the tension in the light, rigid tow bar is T newtons. The normal reaction of the track on the car is N_1 newtons and on the trailer is N_2 newtons. The horizontal resistive force on the car is R_1 newtons and on the trailer is R_2 newtons.

(i) Show the forces acting on the car and trailer.

(ii) What can you say about the size of N_1 and the size of N_2?

If the car is moving at a constant speed,

(iii) write down an equation for the system, considering horizontal forces

(iv) write down an equation for the car, considering horizontal forces

(v) write down an equation for the trailer, considering horizontal forces

If the car is moving with an acceleration of a m s^{-2}

(vi) write down the equation of motion for the system ($F = ma$)

(vii) write down the equation of motion for the car

(viii) write down the equation of motion for the trailer

6.

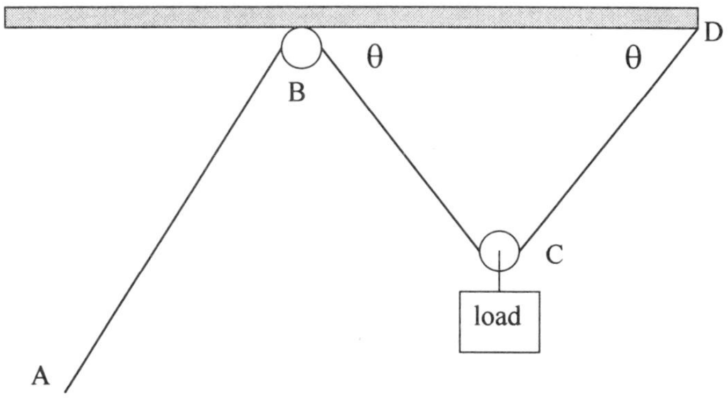

A load is being held above the ground by a man pulling on a rope at A which passes over smooth pulleys on the roof at B and on the load at C and is fixed to the roof at D.

(i) If the man pulls on the rope with a force T newtons, explain why the tensions in the sections of the rope BC and CD are both T newtons.

(ii) Draw a diagram of the forces on the load, which has a mass of M kg.

(iii) Write down an equation for the vertical equilibrium of the load. What happens when you consider horizontal equilibrium?

(iv) The man decides to move closer to B, and he takes in the rope as he moves so that the position of the load does not change. What happens to the tension in the rope AB? Explain your answer.

(i) If AB is at 50° to the line of the horizontal roof, $\theta = 40^\circ$ and the tension in the rope AB is 80 N, find the magnitude and direction of the force on the pulley at B due to the tension in the rope.

GENERAL MOTION OF A PARTICLE

If position, velocity or acceleration of a particle is given to you in terms of time, e.g. displacement, $x = 3t^2 + 2t - 5$ or $\mathbf{r} = 2t\mathbf{i} + (t^2 - 1)\mathbf{j} - 3\mathbf{k}$ or velocity, $\mathbf{v} = 2t^2\mathbf{i} + (5t - 1)\mathbf{j} - 3t\mathbf{k}$, then use:

velocity, $\mathbf{v} = \dfrac{dr}{dt}$

r is the position vector of the particle. Differentiate **r** with respect to t.
e.g. If $\mathbf{r} = t^2\mathbf{i} + 2t\mathbf{j}$ then $\mathbf{v} = \dfrac{dr}{dt} = 2t\mathbf{i} + 2\mathbf{j}$

Note: $\mathbf{v} = \dot{\mathbf{r}}$

acceleration, $a = \dfrac{dv}{dt}$

Differentiate **v** with respect to t.
e.g. If $\mathbf{v} = 3t^2\mathbf{i} + 5\mathbf{j}$ then $a = \dfrac{dv}{dt} = 6t\mathbf{i}$

Note: $a = \dot{\mathbf{v}} = \ddot{r}$

and

velocity, $\mathbf{v} = \int a\, dt$

Integrate a with respect to t.
e.g. If $a = 2t\mathbf{j}$, then, $\mathbf{v} = \int 2t\,\mathbf{j}\, dt$

position, $\mathbf{r} = \int v\, dt$

Integrate **v** with respect to t.
e.g. If $\mathbf{v} = t\mathbf{i} + t^2\mathbf{j}$, then, $\mathbf{r} = \int t\mathbf{i} + t^2\mathbf{j}\, dt$

DON'T FORGET THE CONSTANTS OF INTEGRATION.

1. COMPLETE:
   If displacement, $s = 5t^2 - 4t + 3$

   velocity, $v = \dfrac{ds}{dt} =$ — *Differentiate.*

   acceleration, $a = \dfrac{dv}{dt} =$ — *Differentiate.*

2. COMPLETE:
   If acceleration, $a = 5t - 3$ and $s = 0$ and $v = 5$ when $t = 0$

   velocity, $v = \int a\, dt = \int \qquad dt$ — *Integrate.*

   $=$ — *Don't forget the constant of integration.*

   $v = 5$ when $t = 0$ $\therefore$ $5 =$ — *Find the constant -use the information given.*

   $\therefore$ $v =$ — *Write the formula for v with the constant.*

   displacement, $s = \int v\, dt = \int \qquad dt$ — *Integration.*

   $=$ — *Don't forget the constant of integration.*

   $s = 0$ when $t = 0$ $\therefore$ $0 =$ — *Find the constant -use the information given.*

   $\therefore$ $s =$ — *Write the formula for s with the constant.*

3. COMPLETE: If position, $\mathbf{r} = (2t^3 + 6t)\,\mathbf{i} + (t^2 - 4t)\,\mathbf{j}$ velocity, $\mathbf{v} = \frac{dr}{dt} =$ acceleration, $\mathbf{a} = \frac{dv}{dt} =$	*Differentiate the t s, leave the **i** and **j*** ***(they just tell you the direction)*** *Differentiate.*
4. COMPLETE: If acceleration, $\mathbf{a} = 5t\mathbf{i} - 3\mathbf{j}$ and $\mathbf{r} = 3\mathbf{i}$ and $\mathbf{v} = 5\mathbf{i} + \mathbf{j}$ when $t = 0$ velocity, $\mathbf{v} = \int \boldsymbol{a}\, dt = \int \qquad dt$ $=$ $\mathbf{v} = 5\mathbf{i} + \mathbf{j}$ when $t = 0 \quad \therefore \quad 5\mathbf{i} + \mathbf{j} =$ $\mathbf{v} =$ position, $\mathbf{r} = \int \boldsymbol{v}\, dt = \int \qquad dt$ $=$ $\mathbf{r} = 3\mathbf{i}$ when $t = 0 \quad \therefore \quad 3\mathbf{i} =$ $\therefore \quad \mathbf{r} =$	*Integrate.* *Don't forget the constant of integration.* *Find the constant -use the information given.* *Write the formula for **v** with the constant.* *Integration.* *Don't forget the constant of integration.* *Find the constant -use the information given.* *Write the formula for **r** with the constant.*
5. If $\mathbf{r} = 5t^2\,\mathbf{i} + 8t\,\mathbf{j}$ find $\mathbf{v}$ when $t = 3$. $\mathbf{v} = \frac{dr}{dt} =$ when $t = 3$ $\mathbf{v} =$	*Differentiate the t s in front of the **i** and write down **i** then differentiate the t s in front of the **j** and write down **j**. Substitute for t.*
6. If $\mathbf{r} = 2t^3\,\mathbf{i} + 3t\,\mathbf{j}$ find the speed of the body when $t = 1$ $\mathbf{v} = \frac{dr}{dt} =$ when $t = 1$ $\mathbf{v} =$ speed $= \sqrt{\qquad\qquad} =$ m s^{-1}	*As before differentiate to find the velocity vector when t = 1.* <u>*Speed is the magnitude of this vector.*</u> *Answer to 3 sig. fig.*
7. If $\mathbf{r} = (t^3 - t^2)\mathbf{i} + (3 - 5t^2)\mathbf{j}$ find $\boldsymbol{a}$ when $t = 2$.	*Differentiate to find **v** and then differentiate again to find **a**.*
8. If $\mathbf{v} = 3t^2\,\mathbf{i} + 9\,\mathbf{j}$ find $\boldsymbol{a}$ when $t = 1$.	*Differentiate to find **a**.*
9. If $\mathbf{v} = 6\,\mathbf{i} + 2t\,\mathbf{j}$ find $\mathbf{r}$ when $t = 2$ given that $\mathbf{r} = 2\mathbf{i} + 3\mathbf{j}$ when $t = 0$.	*Integrate. Don't forget the constant of integration.*

10. If $\mathbf{v} = 3t\,\mathbf{i} - 6t^2\,\mathbf{j}$ and $\mathbf{r} = (2\mathbf{i} + \mathbf{j})$ m when $t = 0$, find the distance from the start when $t = 2$.	*Integrate to find the position vector when $t = 2$.* *Then find the magnitude of this vector for distance.*
11. If $\boldsymbol{a} = 2t\,\mathbf{i} - \mathbf{j}$ and $\mathbf{r} = 0$ and $\mathbf{v} = 4\mathbf{i} + 5\mathbf{j}$ when $t = 0$, find $\mathbf{r}$ at time t.	Integrate - don't forget the c. Use $\mathbf{v} = 4\mathbf{i} + 5\mathbf{j}$ when $t = 0$ to find c. Integrate again to find $\mathbf{r}$ - don't forget the constant. Use $\mathbf{r} = 0$ when $t = 0$ to find the constant. Write down $\mathbf{r}$ in terms of t.
12. A mouse, M, moves along a straight track passing through a point O at time, $t = 0$ seconds. The velocity v m s^{-1} of M is given by $v = 4t - t^2$. Find the acceleration of M at time t. Find the maximum velocity, and state at which time it occurs. Calculate the distance travelled in the first 3 seconds. Calculate the time at which the mouse returns to the point O	*DO NOT TRY TO USE THE CONSTANT ACCELERATION EQUATIONS! THIS IS **NOT** CONSTANT ACCELERATION* *Differentiate.* $\frac{dv}{dt}$ *Find t when* $\frac{dv}{dt} = 0$ *Integrate and find the constant of integration (s=0 when t=0) OR put in limits 0 and 3 and integrate* *Use s = …… and find t when s = 0 (one value will be t = 0)*
13. A small remote controlled car moves along a straight line, passing through a point A with velocity 7 m s^{-1} at time $t = 0$. The acceleration of the car after t seconds is $(20 - 6t)$ m s^{-2}. Find the velocity, v m s^{-1} of the car at time t seconds. Show that $v = 0$ when $t = 7$ and find the greatest speed of the car in the interval $0 \le t \le 7$. Find the distance travelled by the car over the same interval	*DO NOT TRY TO USE THE CONSTANT ACCELERATION EQUATIONS! THIS IS **NOT** CONSTANT ACCELERATION.*

14. At time t seconds the acceleration of a cheetah moving in a straight line across a horizontal plane is $(2t - 8)\ m\ s^{-2}$. Given that the initial velocity of the cheetah is $12\ m\ s^{-1}$, find the times when the cheetah comes to instantaneous rest.

*DO NOT TRY TO USE THE CONSTANT ACCELERATION EQUATIONS! THIS IS **NOT** CONSTANT ACCELERATION.*

15. A toy moves in a straight line such that its velocity, $v\ m\ s^{-1}$, is given by

$v = 10t - 2t^2 \qquad 0 \le t \le 3$

$v = 3t - 3 \qquad 3 \le t \le 7$

When t = 0 the toy is at point O. Calculate the displacement of the toy from O when (a) $t = 3$ (b) $t = 5$.

A sketch of the function may help.

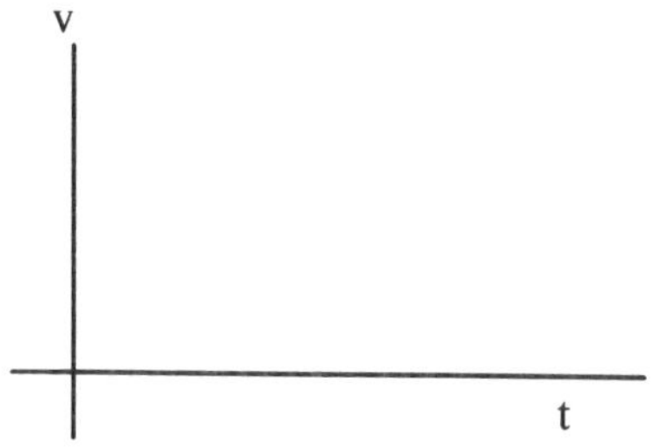

Area under graph equals the displacement.

16. A particle moves along the *x* axis. Its velocity, $v\ m\ s^{-1}$, t seconds after leaving the origin O is given by $v = 12 + 4t - t^2$

Find the acceleration of the particle when its velocity is zero.

Find the distance of P from O when its acceleration is zero.

Calculate the total distance travelled by P in the interval $0 \le t \le 9$.

A sketch of the function may help.

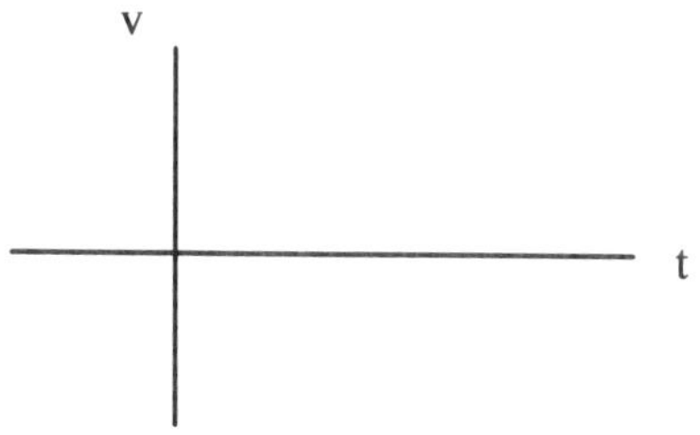

17. A bird flies so that its position vector at time t seconds is given by $\mathbf{r} = (t^2 - 4t)\,\mathbf{i} + (t^3 + k\,t^2)\,\mathbf{j}$ where k is a constant. At t = 0 the bird is at the origin. Find an expression for the velocity of the bird at time t seconds. Given that the bird comes to instantaneous rest, by landing on a branch , find k. Find the distance of the landing point from the origin.	*Distance is the magnitude of the displacement.*
18. A remote controlled car is initially at the origin O. It moves so that at time t seconds its position vector relative to O is given by $\mathbf{r} = (t^3 - 3t)\mathbf{i} + (4t^2 - 2)\mathbf{j}$, $t \geq 0$. Find the velocity of the car at time t seconds. Hence calculate the time at which the car is moving in a direction which is parallel to the vector $\mathbf{i} + \mathbf{j}$.	

In three dimensions you just have another unit vector.

19. If $\mathbf{r} = 4t\,\mathbf{i} - (t^2 - 5t)\,\mathbf{j} + (t^3 - 7)\mathbf{k}$ find **v** and ***a*** when t = 2.	*Differentiate the t s in front of the **i** and write down **i** then differentiate the t s in front of the **j** and write down **j** then differentiate the t s in front of the **k** and write down **k**. Substitute for t.*

If you are given force as a function of time e.g. **F** = 2t**i** + 2**j** and you are asked to find position or velocity or acceleration at some time , then you use **F** = m ***a*** to find acceleration ***a*** (divide by m) and proceed as before.

If you are given position or velocity or acceleration in terms of t and you are asked to find a force, first find the acceleration and then use **F** = m ***a*** to find the force (multiply by m).

1. If the velocity of a particle of mass 5 kg is given by $\mathbf{v} = 3t^2\,\mathbf{i} + 9t\,\mathbf{j}$ find the force acting on the particle when t = 1.	*Differentiate to find **a**.* *Use **F** = ma*
2. If the force acting on a particle of mass 3 kg is given by $\mathbf{F} = 6t\,\mathbf{i} - 9\mathbf{j}$ and the velocity of the particle is 2**i** + 5**j** when t = 0, find **v** at time t.	*Use **F** = ma* Integrate - don't forget the c. Use **v** = 4**i** + 5**j** when t = 0 to find c. Write down **v** in terms of t.
3. The position vector of a bike of mass 10 kg on a track is modelled by $\mathbf{r} = \frac{1}{12}\{(6t^3 - 5t)\mathbf{i} + (t^4 + 2t^3)\mathbf{j}\}$ m. Find the magnitude and direction of the resultant force acting on the bike at t = 3 seconds.	

PROJECTILE MOTION

A particle sent moving into the air is a projectile. Air resistance is ignored and so the only force acting on the particle is its own weight.

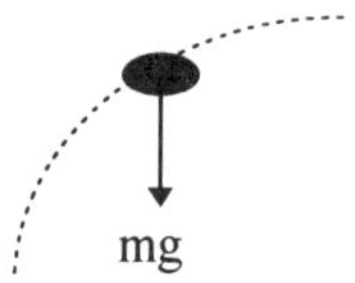

The particle has a downward acceleration of magnitude g ms^{-2} but has constant velocity horizontally.

So you need to analyse the horizontal motion and the vertical motion separately.

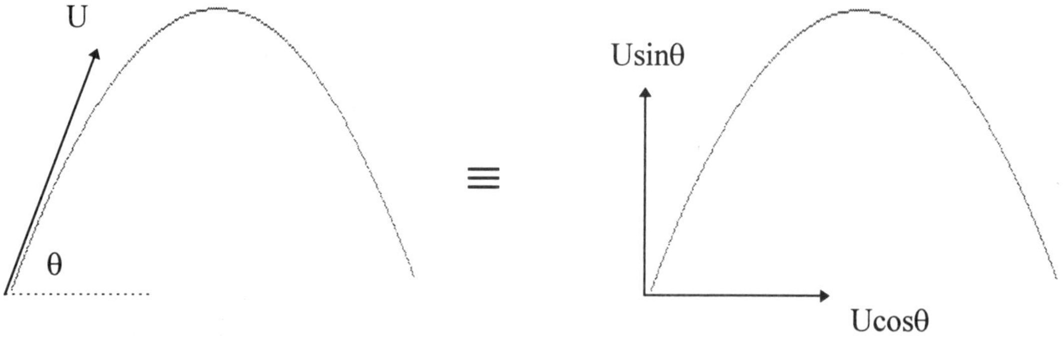

Consider a particle projected with velocity U at an angle θ to the horizontal.
We take the origin as the point of projection, the x axis as the horizontal and the y axis as the vertical.

<u>Horizontally</u>:

Initial velocity $\boxed{U_H = U\cos\theta}$

Displacement x NO ACCELERATION So 's = u t'

so in other words $\boxed{x = U\cos\theta\, t}$

<u>Vertically</u>:

	Initial velocity	$U_V = U\sin\theta$
	Displacement	y
	Acceleration	g ms^{-2} = 9.8 ms^{-2} downwards (in most problems)
	Velocity in flight	V_v

Taking up as positive

Acceleration = -g Initial Vertical velocity $\boxed{U_V = U\sin\theta}$

So $v = u + at \Rightarrow$ $\boxed{V_v = U\sin\theta - g\,t}$

$v^2 = u^2 + 2as \Rightarrow$ $\boxed{V_v^2 = (U\sin\theta)^2 - 2gy}$

$s = ut + \frac{1}{2}at^2 \Rightarrow$ $\boxed{y = U\sin\theta\, t - \frac{1}{2}g\,t^2}$

These 6 highlighted equations are all that are needed to solve problems on the motion of any projectile.

You do not need to know all these -
you can just use:
the constant acceleration equations and initial velocity, $U_V = U\sin\theta$ VERTICALLY
and s = u t and initial velocity, $U_H = U\cos\theta$ HORIZONTALLY

1. A projectile is thrown as shown. Complete the calculations.

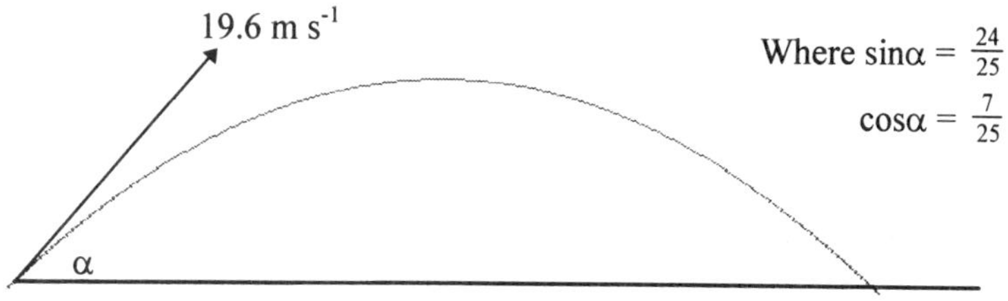

Where $\sin\alpha = \frac{24}{25}$

$\cos\alpha = \frac{7}{25}$

TAKING UP AS POSITIVE.

Horizontal component of initial velocity, $U_H =$ $=$

Vertical component of initial velocity, $U_V =$ $=$

(a)Time to greatest height

Vertical component of velocity will be zero.

Vertically: $U_V =$ $V_V =$ $a =$

Using '$v = u + at$'

$0 =$

$t =$

(b) Time to hit ground

The time of flight.
Here the displacement vertically will be zero.

Vertically: $U_V =$ $s =$ $a =$

Using '$s = ut + \frac{1}{2}at^2$'

$0 =$

$=$

Factorise.

$t =$ or $t =$

$\therefore$ time to hit ground is

(c) Horizontal range

Horizontally: $U_H =$ $t =$

Use the time from part (b)

Using '$s = ut$'

$x =$ $=$ i.e the horizontal range.

(d) The greatest height

Vertically: $U_V =$ $V_V =$ $a =$

Using '$v^2 = u^2 + 2as$'

Vertical component of velocity will be zero.

$0 =$

$s =$ $=$ i.e the greatest height

(e) Speed of projectile at time t = 1.

Vertically: $U_V =$ $a =$ $t = 1$

using '$v = u + at$'

Speed is the magnitude of the velocity.
Find the components of velocity at t = 1 and then use Pythagoras.

$V_V =$

Horizontally: $V_H = U_H =$

Horizontally there is no acceleration so the component stays the same.

$\therefore$ Speed $= \sqrt{\qquad\qquad}$ $=$

2. A projectile is launched from the ground with an initial velocity of 22 m s^{-1} at an angle of θ° to the horizontal, where $\tan\theta^\circ = \frac{3}{4}$.

(i) Write down the position of the projectile after time t.

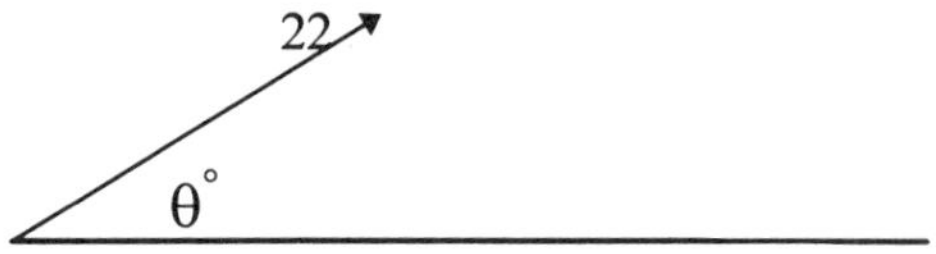

Horizontally:

Vertically:

(ii) Show that the equation of the path is the parabola

$$y = 0.75x - 0.0158x^2$$

(iii) Find y when $x = 4$

(iv) Does the projectile ever rise above 9 m above the ground?

TAKING UP AS POSITIVE

Sketch the situation first.

Then write down the horizontal and vertical components of the initial velocity (or draw them on the diagram).

'$s = ut$'

'$s = ut + \frac{1}{2}at^2$'

Find $t =$ from the first equation

Substitute for t in the second equation.

Substitute $x = 4$ into $y = 0.75x - 0.0158x^2$

Substitute $y = 9$ into $y = 0.75x - 0.0158x^2$ *and see if you can work out the values of x for which $y = 9$.*

3. A ball is thrown horizontally with a speed of 14 ms^{-1} from a window 10 m above the ground. Find the time taken before the ball hits the ground. Find also the horizontal distance from the window when the ball hits the ground.

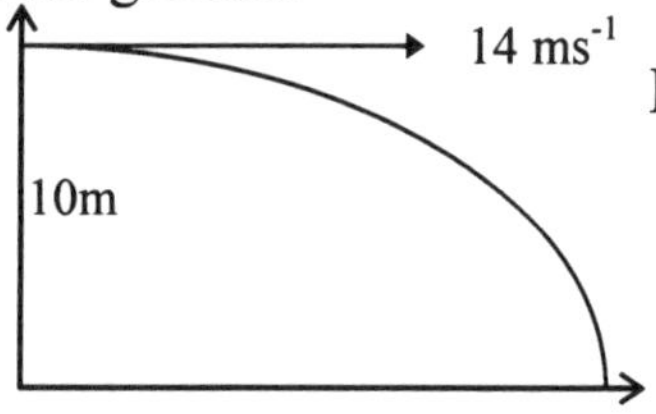

Horizontally:		Vertically:	
u =		u =	
s =		s =	
t =		a =	
a =		t =	

Take RIGHT on the diagram as positive
Take DOWN as positive

Vertical motion: $s = ut + \frac{1}{2}at^2$

$=$ $\Rightarrow$ $= t^2$

$t =$ $\Rightarrow t =$ secs (3 s.f.)

Horizontal motion: $s = ut$

$x =$ $=$ $=$ m

For this example ALL motion is to the RIGHT and DOWN - so these are our positive directions

Horizontally the only useful equation is s=ut. You do not know x or t, so move onto considering vertical motion.

You know s, u, and a and want t. Use $s = ut + \frac{1}{2}at^2$

Give your answer for t to 3 s.f....

... but when you need the value of t later use the full calculator value for t for most accuracy.

If you want the speed of the particle at a time t secs. you want the length of the velocity vector.

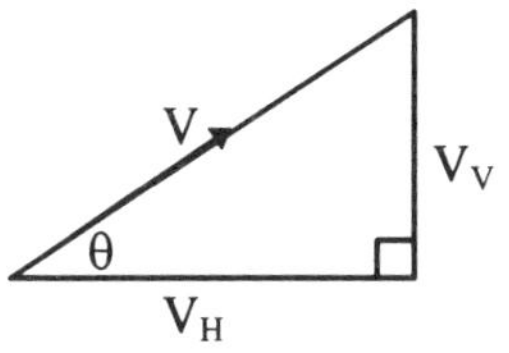

Use Pythagoras: Speed = length of V = $\sqrt{v_V^{\ 2} + v_H^{\ 2}}$

If you want the direction of motion you need to find the angle between the velocity vector and the horizontal, θ in the diagram.

Tan θ = $\dfrac{V_V}{V_H}$

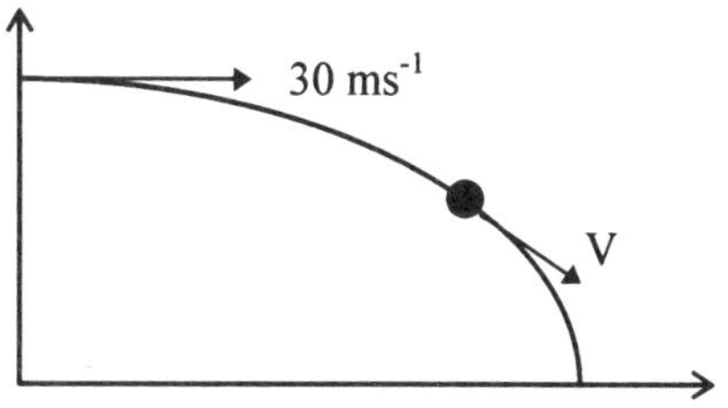

4. A tennis ball is hit horizontally with a velocity of 30 ms^{-1}. Find the speed and direction of the ball 0.5 seconds after projection. Horizontally: u = 30 v = u t = 0.5 Vertically: u = 0 v = ? a = 9.8 t = 0.5 Vertically: v = u + at v = 0 + 9.8 × 0.5 = Horizontally : constant vel^y 30 m s^{-1}	*TAKING DOWN AS POSITIVE.* *Horizontally there is no acceleration, so velocity is constant horizontally* *Sketch the velocity triangle. Horizontal 'followed by' vertical.* *Use Pythagoras and tan to find the speed and direction.*
5. A particle is projected with velocity 15 m s^{-1} at an angle of 60° to the horizontal. Calculate the greatest height it reaches above the horizontal through the point of projection.	*Sketch.* *Write down the horizontal and vertical components of initial velocity.* *At greatest height, the vertical component of velocity is zero.* *Use $v^2 = u^2 + 2$ as vertically.*

6. A man standing on the edge of a vertical cliff 50 m above the sea throws a stone at 30 m s^{-1} out to sea at an angle of 30° above the horizontal. Calculate:

   (a) the time the stone takes to reach the sea.
   (b) the horizontal distance of the man from the point where the stone hits the sea.

7. A body is projected from a point on a horizontal plane, with a velocity 40 m s^{-1}, at an angle of 60° to the horizontal. The point of projection is at a horizontal distance of 40 m from the foot of a vertical mast which is 30 m high, and the motion takes place in a plane perpendicular to the wall. Calculate

   (i) the vertical height by which the body clears the wall,
   (ii) the greatest height above the horizontal plane reached by the body,
   (iii) the time of flight of the body,
   (iv) the horizontal distance beyond the wall at which the body strikes the plane.

8. A tower 30 m high stands on a horizontal plane. From the top of the tower a ball is thrown with a velocity of 25 m s^{-1} at an angle of 30° above the horizontal. Calculate

(i) the greatest height above the plane reached by the ball
(ii) the time taken by the ball to reach the plane
(iii) the horizontal distance travelled by the ball, before it hits the plane
(iv) the speed of the ball on striking the plane.

9. A gunman fires a gun horizontally with the barrel resting on a wall 2m high. The bullet hits the ground 300 m from the wall. Find the initial speed of the bullet.

Exam type questions

1. A goods train consists of a locomotive of mass 30 tonnes and two trucks each of mass 15 tonnes. The locomotive and the trucks are linked by light rigid couplings and it is moving on a straight horizontal track. The locomotive is producing a driving force from the engine of 20 kN forwards and the resistances to motion are 1 kN for each truck and 3 kN for the locomotive.
   (i) Show that the acceleration of the whole train is 0.25 m s^{-2}.
   (ii) Draw a diagram of the horizontal forces acting on the rear most truck. Calculate the tension in the rearmost coupling.
   (iii) Find the tension in the front coupling.
   The locomotive shuts off its engine and applies a braking force of 4 kN.
   (iv) Find the deceleration of the whole train.
   (v) Find the forces in each coupling and determine if each is in tension or compression.

2. (a) A firework explosive is projected vertically upwards and reaches a maximum height of 90 m above its point of projection. Find the speed of projection. (You may assume that the only force acting after projection is its weight and it acts as a particle.)

   (b) Another firework explosive is projected from ground level at an angle of 60° to the horizontal, and it explodes after two seconds at the top of its trajectory, 40 m horizontally from its point of projection.

   (i) Find the initial speed of projection. (Same assumption as above)

   (ii) How high above the ground does the explosion take place?

   (c) A third firework explosive is projected from ground level at $50\ \text{m s}^{-1}$ at an angle of 60° to the horizontal, and fails to explode. Show that the rocket is more than 30 m above the ground for about 7.3 seconds.

   (d) A fourth firework explosive is to be projected at $50\ \text{m s}^{-1}$ to explode after 3 seconds, 100m above the ground. At what angle should it be projected?

3. A dry ski slope centre uses a series of rope winches to transport skiers from the end of the slope back to the top. First the skier is pulled along a horizontal track, then along a track at an angle of 30° to the horizontal. You may assume that the pulling rope is light and inelastic and remains parallel to the direction of motion of the skier.

   A skier of mass 80 kg is being pulled along the horizontal track, against a constant resistance to motion of 40 N

   (i) Initially the skier is moving at a constant speed. Calculate the tension in the pulling rope.

   (ii) Later the skier has an acceleration of 0.3 m s^{-2}. Calculate the new tension in the pulling rope.

   The skier then moves on to the sloping track. The resistance to motion continues to be 40 N

   (iii) Draw a diagram to show the forces acting on the skier

   (iv) If the skier moves at a constant speed up the track, find the tension in the rope.

   (v) As the skier moves towards the end of the track she decelerates at 0.2 m s^{-2}. Calculate the new tension in the rope.

   (vi) This final stage of the journey when the deceleration is 0.2 m s^{-2} lasts for 10 seconds, and the skier comes to rest. How far does the skier travel over this final stage?

4. A mechanical arm picks up an object from the end of a conveyor and moves it to another conveyor at floor level. A simple model for the position of the object at time t is given by
$\mathbf{r} = (4.5t^2 - t^3)\mathbf{i} + (9 - t^2)\mathbf{j}$ for $0 \le t \le 3$.
The unit vectors **i** and **j** have directions which are horizontal and vertical respectively and the origin is a point on the floor, vertically below the object when t = 0. The unit of length is the metre.
(i) How far above the floor is the object initially?
(ii) Show that the object is at floor level when t = 3, and find its horizontal distance from O at that time.
(iii) Find expressions for the velocity and acceleration of the object at time t.
(iv) At what speed is the object travelling when t = 1.
(v) Show that the magnitude of the acceleration of the object when t = 2 is about 3.61 m s^{-2}. Calculate the magnitude of the resultant force on the object at this point if its mass is 20 kg.
(vi) Does the suitcase move only vertically at any stage? Explain your answer.

5.

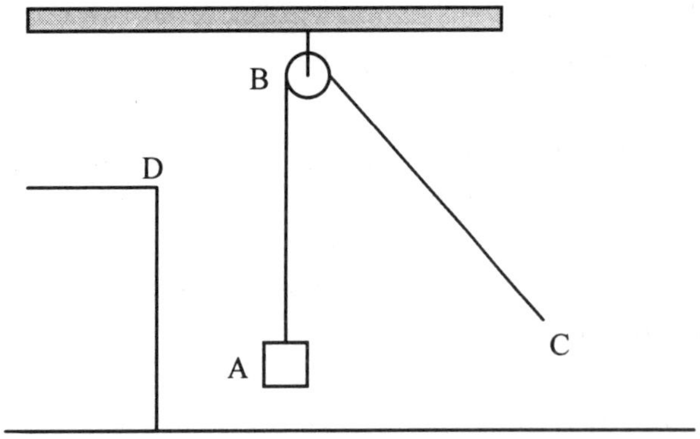

A worker at C supports a bucket of cement, A, of mass 50 kg by a rope passing over a smooth pulley, B.
The bucket is stationary.
(i) Draw a diagram showing the forces acting on the bucket.
(ii) Write down the tension in AB. What is the tension in BC?

The man wants to lift the bucket up to his colleague at D and he pulls on the rope with a constant force of 500 N.
(iii) Find the acceleration of the bucket.
(iv) As the bucket reaches D it is moving at 1 m s^{-1}. How far above A is D?
At this moment the man is distracted and lets go of the rope.
(v) If the bucket is 4 m above the ground , how long will it take for the bucket to hit the ground?
Later in the day the man raises a bucket from rest through 2 m in 4 seconds, and then through another 3 m at which stage the bucket is moving at 2 m s^{-1}.
(vi) Does this evidence suggest the man is pulling with a constant force?

NOW TRY SOME PAST EXAMINATION PAPERS.

6. A car of mass 1500 kg tows a caravan of mass 500 kg. The car and the caravan move on a straight road with the engine of the car exerting a constant pull of 4100 N. Given that there are frictional forces of 800 N on the car and 300 N on the caravan, find the magnitude of
   (a) the acceleration of the car
   (b) the tension in the tow bar between the car and the caravan.
   The car and the caravan then go up a straight road which is inclined at an angle θ to the horizontal where $\sin\theta = \frac{1}{10}$, the frictional resistances on the car and caravan remaining unchanged.
   (c) Draw a diagram showing all the forces acting on the car and the caravan.
   The speed of the car and caravan increases from 10 m s^{-1} to 20 m s^{-1} with the engine of the car now exerting a constant pull of P newtons.
   (d) Find the value of P.
   When the speed is 20 m s^{-1} the tow bar breaks.
   (e) Show that the deceleration of the caravan is 1.58 m s^{-2}
   (f) Hence, find the further distance (to the nearest metre) travelled by the caravan before it comes to instantaneous rest.

Answers

<u>Page4</u>

Assumptions: light, inextensible rope.
Smooth pulley
The boat can be modelled as a particle
The boat, chute, rope and pulley are in the same vertical plane

<u>Page6</u>

$2\underline{b} = 2 \times \begin{bmatrix} 5 \\ 1 \end{bmatrix} = \begin{bmatrix} 10 \\ 2 \end{bmatrix}$

$2\underline{b} = 2(5\mathbf{i} + \mathbf{j}) = 10\mathbf{i} + 2\mathbf{j}$

$\underline{a} - \underline{b} = \begin{bmatrix} 3 \\ -4 \end{bmatrix} - \begin{bmatrix} 5 \\ 1 \end{bmatrix} = \begin{bmatrix} -2 \\ -5 \end{bmatrix}$

$\underline{a} - \underline{b} = -2\mathbf{i} - 5\mathbf{j}$

$2(\underline{a} + \underline{c}) = 2\times\left(\begin{bmatrix} 3 \\ -4 \end{bmatrix} + \begin{bmatrix} 0 \\ -2 \end{bmatrix}\right) = \begin{bmatrix} 6 \\ -12 \end{bmatrix}$

$2(\underline{a} + \underline{c}) = 6\mathbf{i} - 12\mathbf{j}$

$2\underline{a} + \underline{b} - 3\underline{c}$

$= 2 \times \begin{bmatrix} 3 \\ -4 \end{bmatrix} + \begin{bmatrix} 5 \\ 1 \end{bmatrix} - 3\times\begin{bmatrix} 0 \\ -2 \end{bmatrix} = \begin{bmatrix} 11 \\ -1 \end{bmatrix}$

$2\underline{a} + \underline{b} - 3\underline{c} = 11\mathbf{i} - \mathbf{j}$

$|2\underline{b}| = \sqrt{104}$ 11.3° to the **i** direction

$|\underline{a} - \underline{b}| = \sqrt{29}$ -111.8° to the **i** direction

<u>Page7</u>

(i) **q-p** (ii) ½ (**q-p**) (iii) ½ (**p+q**)
(iv) ½ (**q-p**) (v) ½ **q**

<u>Page8</u>

OP=5**i** + 2**j** **OQ**=7**i** + **j** **OR**=11**i** - **j**
PQ =2**i** - **j** **PR**=6**i** - 3**j**
PR=3**PQ** ∴P,Q and R are on a straight line.

1. (i) $\begin{bmatrix} -2 \\ 7 \end{bmatrix}$ (ii) $\begin{bmatrix} -3 \\ 6 \end{bmatrix}$ (iii) $\begin{bmatrix} 5 \\ 3 \end{bmatrix}$ (iv) $\begin{bmatrix} -3 \\ -3 \end{bmatrix}$

2.

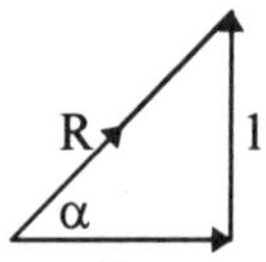

R = √(100 + 36) = 11.7
tanα=10/6 α=59.0° to the 6 N force

3. (i) 20 N in the same direction.
(ii) 4N in the P direction
(iii)

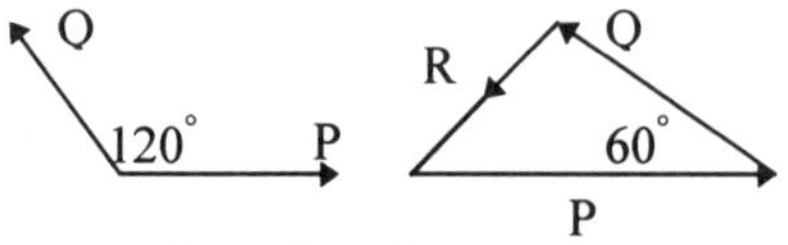

$R^2 = 12^2 + 8^2 - 2\times12\times8\cos60$
$R = \sqrt{112} = 10.58$
cosθ = (112+144-64)/(2×10.58.×12)
θ = 40.9° to the P force

<u>Page9</u>

4.(a)

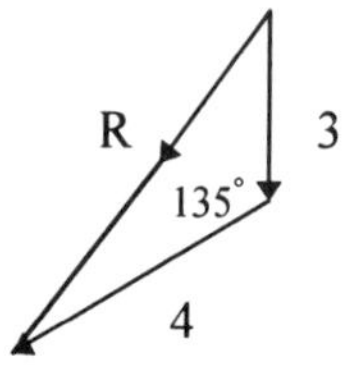

$R^2 = 3^2+4^2-2\times3\times4\cos135$
$R = \sqrt{41.97} = 6.48$
$\cos\theta = (3^2+6.48..^2-4^2)/(2\times3\times6.478..)$
θ = 25.9° bearing 205.9°

(b)

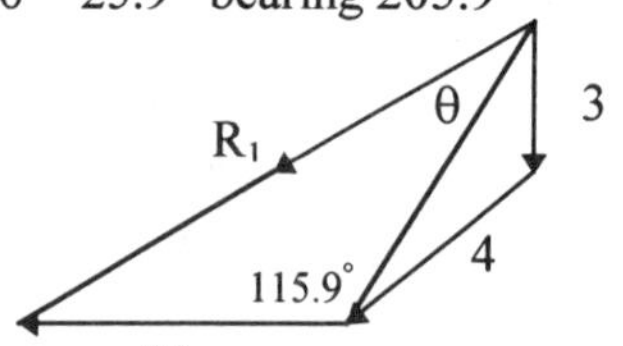

$R_1 = 41.97+100-2\times\sqrt{41.97}\times10\cos115.9$
$R_1 = \sqrt{198.538}$ R=14.09 = 14.1
cosθ = <u>(41.97+198.538-100)</u> / (2×√41.97×√198.538)
θ = 39.6796° = 39.7°
bearing 205.9°+39.7° = 245.6°

1.(i)

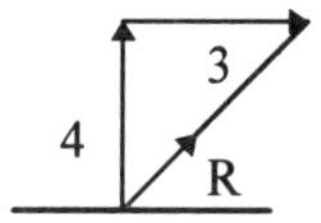

R = √(16 + 9) = 5
tanθ = 3/4 θ = 36.9°
(ii) cos 36.9 = 200/d d = 250 m
t = 250/5 = 50 seconds
(iii) tan36.9 = l/200
l = .75×200 = 150 m

<u>Page10</u>

1. Boat: 2cos60**i** + 2sin60**j**
Current: 1.5cos45**i** - 1.5sin45**j**
Wind: 1cos70**i** + 1sin70**j**
2.40268**i** +1.61108**j**
Speed = $\sqrt{(1.611^2+1.90268^2)}$
= 2.49 m s^{-1}
Direction
tanα=1.611/1.90268=0.84674
α = 40.3° to the E or bearing 049.7°

<u>Page11</u>

Total distance travelled =500 m
Total displacement = 100 m East
Average speed = 500/180
=2.78 m s^{-1}
Average velocity = 100/180
= 0.556 m s^{-1}East

<u>Page12</u>

2.Distance travelled
from A to B = 10
Displacement of B from A =+2
Speed of body at A =5 m s^{-1}
Velocity of body at A = +5 m s^{-1}
Speed of body at B = 3 m s^{-1}
Velocity of body at B = -3 m s^{-1}
3. 40000mh^{-1} = 40000/3600
= 11.1 m s^{-1}
4. 4×3600 = 14400 m h^{-1}
= 14400/1000=14.4 km h^{-1}
5. (i) time = 6/12 = .5hours
(ii) time= 8/20 = .4hours
(iii) Av speed = 14/.9
=15.6 km h^{-1}

<u>Page13</u>

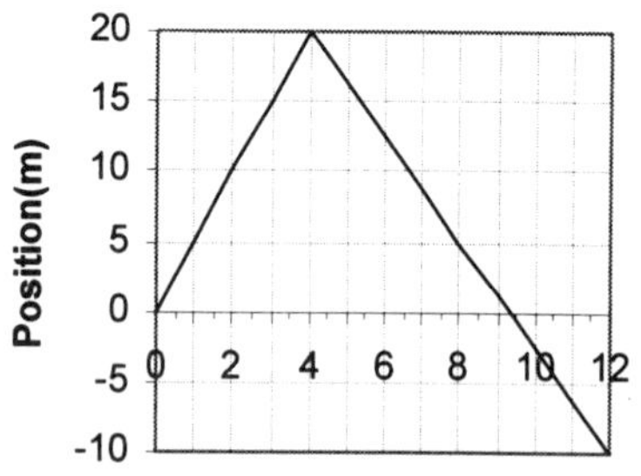

20/4 = +5 m s^{-1} -30/8= -3.75 m s^{-1}

<u>Page14</u>

(ii) First part of motion
Speed = 50/10 = 5ms^{-1}
Second part of motion
Speed = 28/7 = 4 ms^{-1}
Last part of motion
Speed = 36/12 = 3 ms^{-1}
(iii) Average speed = 114/29
= 3.93 ms^{-1}
(v)First part of motion
velocity =50/10 = +5 ms^{-1}
Second part of motion
velocity = -28/7 = -4 ms^{-1}
Last part of motion
velocity = 36/12 = +3 ms^{-1}
(vi) Average velocity = +58/29
= +2 ms^{-1} i.e. forward

<u>Page15</u>

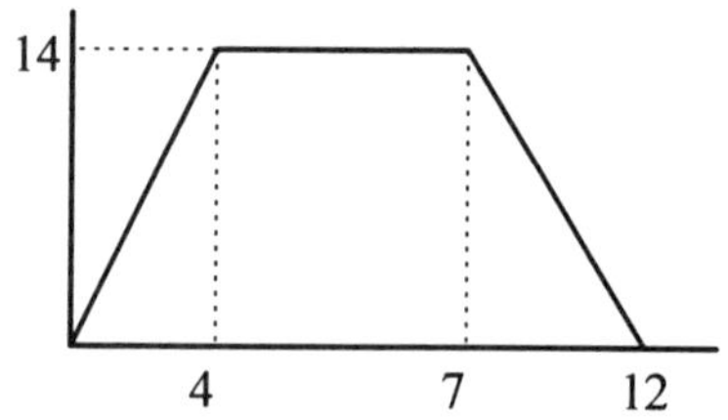

b) 14/4 = 3.5 m s^{-1} -14/5 =-2.8 m s^{-1}
Distance = 14(3+12)/2 = 105 m

<u>Page16</u>

1. u = 2 v = 7 a =1 m s^{-2}
distance = (2+7)5/2+3×7=22.5+21
= 43.5 m
2. u =3 v =-1.5 a =-4.5/6= -0.75
displacement =12+.5×4×3-.5×2×1.5
= 16.5 m in the +ve dir^n
distance = 12+.5×4×3+.5×2×1.5
= 19.5 m

<u>Page17</u>

3.

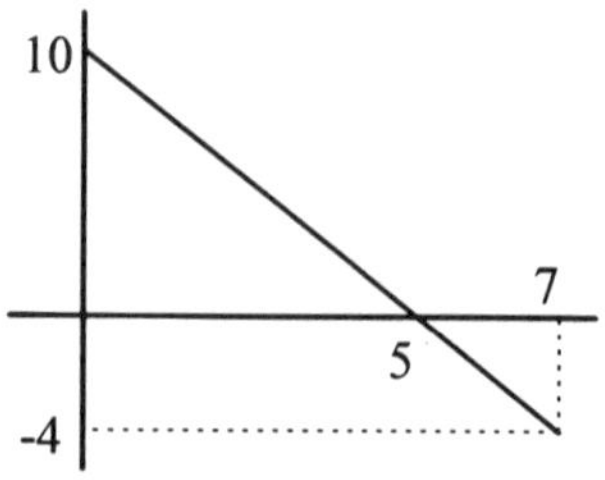

Displacement =.5×5×10-.5×2×4
= 21 m in the +ve dir^n
Distance travelled
=.5×5×10+.5×2×4
= 29 m

4.(i)

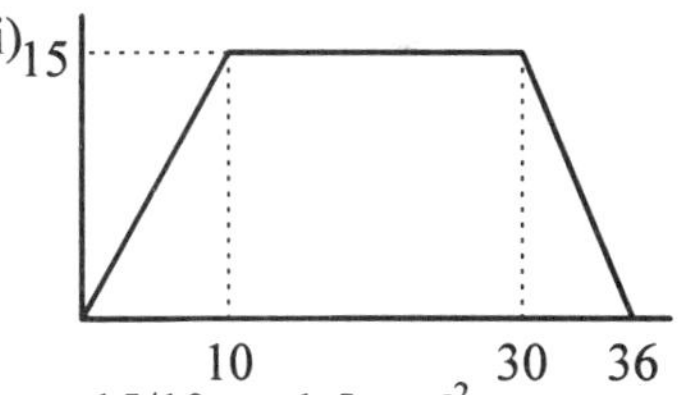

(ii) a = 15/10 = +1.5 m s^{-2}

(iii) $decel^n$ = 15/6 = 2.5 m s^{-2}

(iv).5×10×15+20×15+.5×6×15
= 420 m

Page18

5.(i)

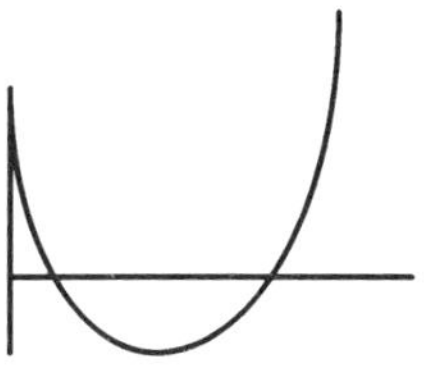

t = 0 1 2 3 4 5 6
s = 3 -1 -3 -3 -1 3 9

(ii) at t = 2.5 s = -3.25
3 + 3.25+3.25 + 9 = 18.5

(iii) 15.25/6 = 2.54 m s^{-1}

(iv) 9/6 = +1.5 m s^{-1}

6.

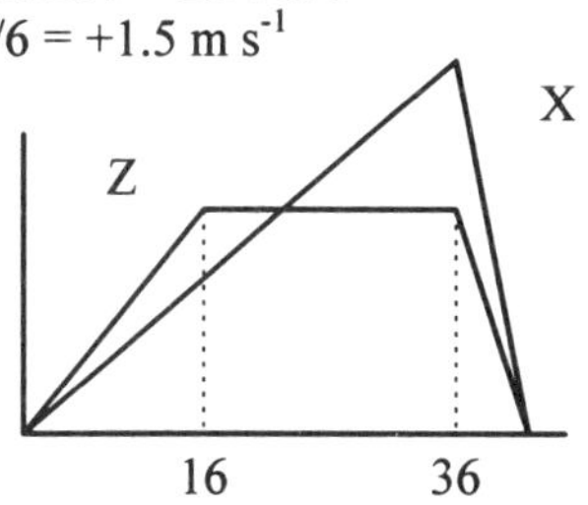

(i) v_z /16 = 1.25
v_z = 16×1.25 = 20 m s^{-1}

(ii) 20/t = 2 t = 10 total time = 46
Distance = (46+20)20/2 = 660 m

(iii) t = 34.5

(iv) 660 = 46×v_x /2 v_x= 28.7 m s^{-1}

Page19

1. (a) s = 0 + ½ ×7×64 = 224
(b) v = 5 + 3×5 = 20
(c) s = 0 + ½ ×2×25 = 25
(d) s = ½ 10(-5+7) = 10
(e) 81 = 49 + 4a a = 8
(f) 32 = u +18 u = 14
(g) 169 = 289 + 20a a = -6
(h) 32 = 4u + 16 u = 4
(i) 144 = 6t + 36 v = 18
(j) s = 50 - 150 = -100
(k) -2 = 6 + 6a a = -4/3

Page20

2. '$v^2 = u^2 + 2as$'
225 = 0 + 60a a = 3.75 m s^{-2}

3. 'v = u +at'
0 = 20 + 6a a = 3.33 m s^{-2}

4. v = 96×1000/3600 = 26.6667
'v = u +at'
26.6667 = 12a a = 2.22 m s^{-2}
'$s = ut + ½ at^2$'
s = 1.111×144 = 160 m

5. '$s = ut + \frac{1}{2}at^2$'
hare: s = 5t + 0
greyhound: s = 0 + ½ $4(t-1)^2$
5t = ½ $4(t-1)^2$
0 = $2t^2$ - 9t + 2
t = 0.234 (not possible) or 4.27 s

6. 'v = u +at'
4 = 1.5 + 10a a = 0.25 m s^{-2}

'$v^2 = u^2 + 2as$'
= 1.5^2 + 2×0.25×200
= 102.25 v = 10.1 m s^{-1}

Page21

1. cuts t axis at 7
(a) 2+2×3 = 8 m s^{-1}
(b) 8 + (3×4) = 20 m s^{-1}
(c) 20 + ½ ×2×4 -½ ×1×2= 23ms^{-1}

Page22

1. (i) t = 1.53
(ii) 0 = 15t - $4.9t^2$
t = 3.06
(iii) t = 3.14 secs
(iv) v = 15.8 m s^{-1}

Page23

2. $4.9T^2 = 15T-15+4.9T^2-9.8T+49$
0 = 5.2T - 10.1 T = 1.942
s = $4.9T^2$ = 18.5 m

3. v^2 = 0 + 39.2 v = 6.26 m s^{-1}

4. '$s = ut + \frac{1}{2}at^2$'
70 = $4.9t^2$ t = 3.78

5. '$v^2 = u^2 + 2as$'
= 0.25 + 490 v = 22.1 m s^{-1}
'$s = ut + \frac{1}{2}at^2$'
25 = 0.5t + $4.9t^2$
0 = $4.9t^2$ + 0.5t - 25
t = 2.21

or 'v = u +at'
22.141 = 0.5 + 9.8t t = 2.21

6. (i)'$s = ut + \frac{1}{2}at^2$'
10 = 20t - $4.9t^2$
$4.9t^2$ - 20t + 10 = 0
t = 3.49825 or 0.58338
time above 10 m is 2.91 secs.
(ii) '$s = ut + \frac{1}{2}at^2$'
8 = 20t - $4.9t^2$
$4.9t^2$ - 20t + 8 = 0
t = 3.632129689 or 0.44950
time above 10 m is 3.18 m

Page24

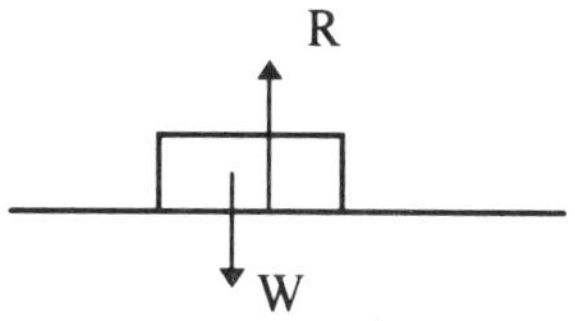

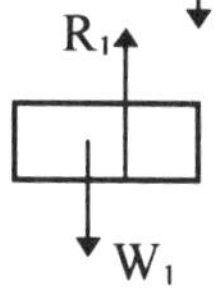

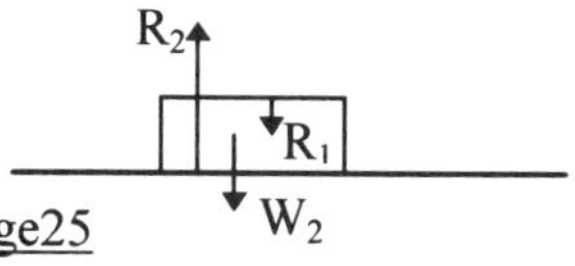

Page25

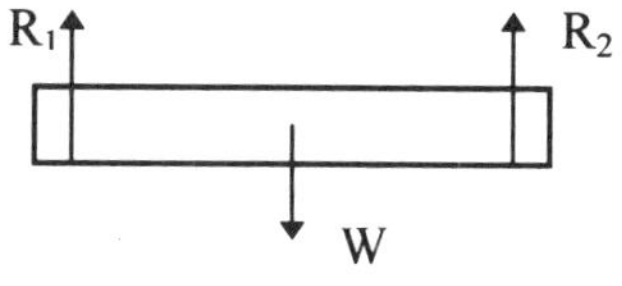

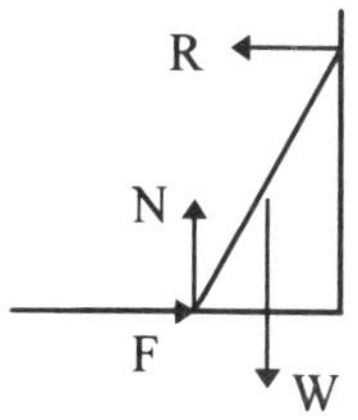

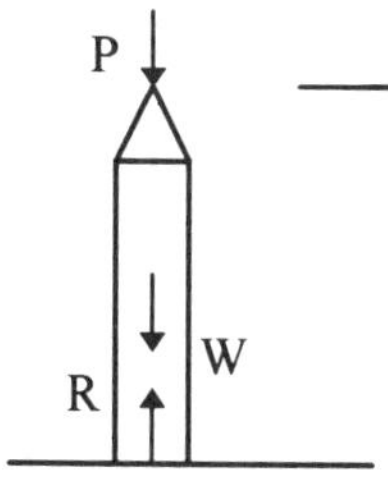

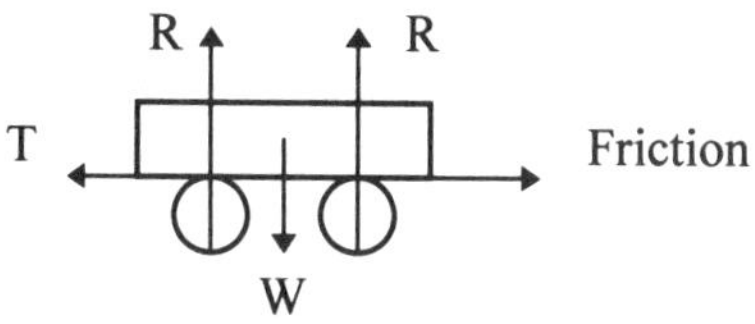

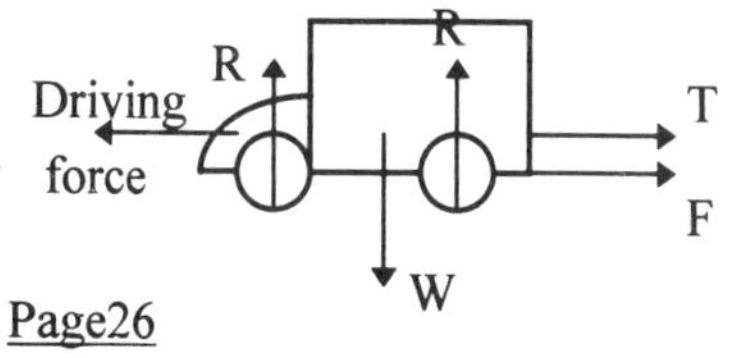

Page26

Horizontal component = 10cos25°

Vertical component: 10sin25°

Page27

1. hor: 15cos20 = 14.1
vert: 15sin20 =5.13

hor: 9cos80 = 1.56
vert: 9sin80 = 8.86

hor: -150cos50 = -96.4
vert: 150sin50 = 115

hor: 20sin35 = 11.5
vert: 20cos35 = 16.4

hor: -80cos43 = -58.5
vert: -80sin43 = -54.6

x dir^n: -120sin24 = -48.8
y dir^n: 120cos24 = 110

x dir^n: -50cos72 = -15.5
y dir^n: -50sin72 = -47.6

x dir^n: 80cos82 = 11.1
y dir^n: -80sin72 = -79.2

Page28

1. 9.43 N at +58.0°to **i** direction
9.22 N at -12.5°to **i** direction
8.60 N at +144.5°to **i** direction
8.54 N at -110.6°to **i** direction
87.7 N at +151.4°to **i** direction

Page29

1. 5**i** - 2**j**
2. (a) 29 N leftwards
(b) 95 N downwards

Page30

1. *x* direction:
80cos40+50cos50-70 = 23.4229359
y direction:
80sin40 - 50sin50 = 13.1207866
Mag: √720.788969 = 26.8
Dir: tanα = 13.12../23.42..
α = +29.3 to the *x* direction

2. x direction:
   45cos30 - 30cos25 = 11.8
   y direction:
   45sin30 + 30sin25 - 80 = -44.8
   11.8**i** - 44.8**j**

Page31

3. 18 N downwards
4. 22 N in the 43 N direction.
5. 5.19**i** + 7.74**j**

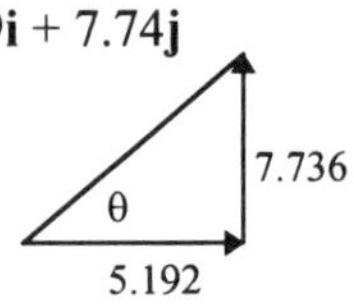

or 9.32 N at 56.1° to the x direction

6. Parallel to 47 N:
   47 - 56sin25 = 23.333377
   Parallel to 64 N:
   64 - 56cos25 = 13.24676

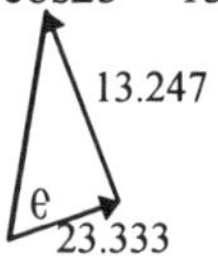

Mag: 26.8 N
Dir: α = 29.6° to the 47 N force

7. Parallel to slope:
   675cos20 - 215 - 400sin30
   = 219 N up the hill

Page32

A = 41 N
X = 1690 N
X = 54 N Y = 62 N

Page33

1. Resolve horizontally:
   F = 140 N
   Resolve vertically:
   N = 20g = 20×9.8 = 196 N
2. R = 45g = 441 N
3. **i** : 2+5-2+p = 0 p = -5
   j : 4-q+5-6 = 0 q = 3

4.(i) x = -7 y = 4
(ii) x = 5 y = 5
(iii) x = 1 y = 2

Page34

x: Pcos30 + Qcos60 =10
0.866P + 0.5Q =10 (i)
y: Psin30 + Qsin60 = 12
0.5P + 0.866Q = 12 (ii)
(i) ×0.5 0.433P + 0.25Q = 5..(iii)
(ii) ×0.866
0.433P + 0.75Q = 10.3923 (iv)
(iii) - (iii) 0.5Q = 5.3923
Q = 10.8
in (ii) 0.5P+ 9.3397 = 12
P = 5.32

Page35

2. Parallel to right hand string(x)
   T_2 - 8gsin30 = 0
   T_2 = 39.2
   Parallel to left hand string(y)
   T_1 - 8gcos30 = 0
   T_1 = 67.9 N

Page36

3. sinθ = 12/20 =3/5
   cosθ = 4/5
   vert: t sinθ = 2g T = 24.5 N
   hor: P = T cosθ = 24.5×0.6 =14.7 N

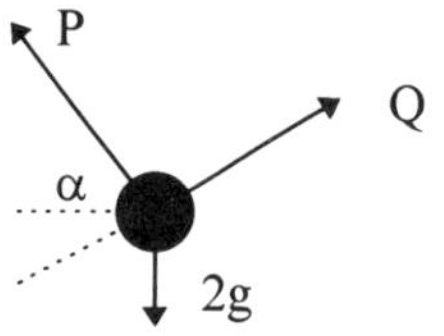

cosα = ½ α = 60°
Perp to string Q = 19.6cosα
= 19.6×0.5 = 9.8 N
Parallel to string: T = 19.6sinα
= 17.0 N

Page37

4.

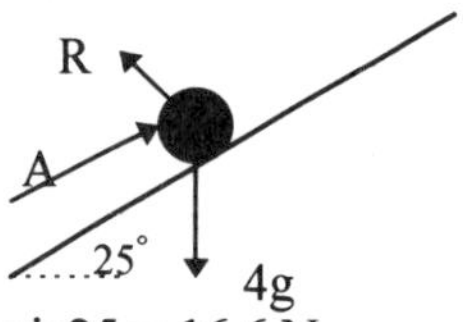

Parallel to slope: A = 4gsin25 = 16.6 N
Perpendicular to slope:
R = 4gcos25 = 35.5 N

5.

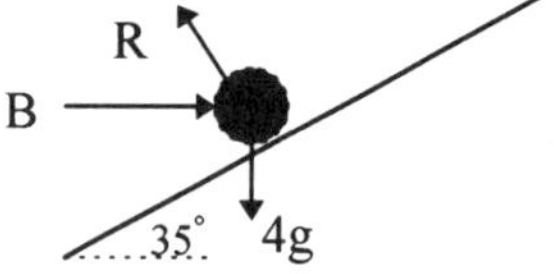

Parallel to slope:Bcos35 = 4gsin35 B = 27.4 N
Perpendicular to slope:
R = Bsin35 + 4gcos35 = 47.9

6.

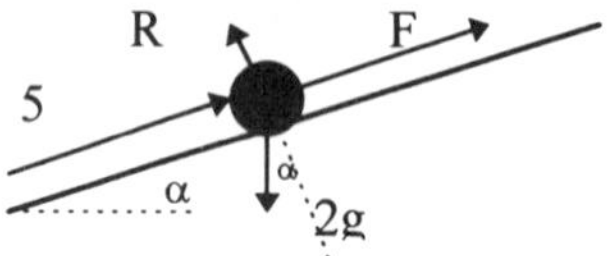

cosα = 12/13 sinα = 5/13
Perpendicular to plane:
R = 2gcosα =18.1 N
Parallel to plane:
F + 5 = 2gsinα
F = 2.54 N

Page38

7. Perpendicular to beach:
   R = 200gcos15 = 1890 N
   Parallel to beach:
   P + 200gsin15 = 700
   P = 193 N
8. Perpendicular to direction of motion:
   Tsin20 = 70sin25
   T = 86.5 N
   Drag = Tcos20 + 70cos25 = 145 N
9. X = 8000cos10 + 7000cos11 = 14700 N
   Y = 8000sin10-7000sin11 = 53.5 N
   14700 N
   53.5 N
   53.5 in the negative y direction
   14700 N

Page39

1. a = 25/6 = 4.17 m s^{-1}
2. F = 850×0.5 = 425 N
3. 'v = u + at'
   8 = 2 + 3a a = 2
   F = 750×2 = 1500 N
4. '$v^2 = u^2 + 2as$'
   0 = 144 + 10a a = -14.4
   F = 550×(-14.4) = -7920 N
   Braking force is 7920 N
5. In the direction of motion:
   550-375 = 400×a a = 0.4375 m s^{-2}
6. 61.5-9.8m = 2.5m
   m = 5 kg
7. 25grams = 0.025 kg
   In direction of motion:
   -1000 = 0.025 a
   a = -40000
   Deceleration 40 000 m s^{-2}

Page40

1. 3**i** + 7**j** = 3***a*** ***a*** = **i** + 7/3 **j**
2. F = 5(5**i** - 2**j**) = 25**i** - 10**j**
   Mag: =$\sqrt{(25^2+10^2)}$ = $\sqrt{725}$ = 26.9N
   tanθ = 10/25 = 2/5 θ =21.8°
   21.8° below **i** direction
   or -21.8° to the **i** direction
3. Vert: R = 3g = 29.4 N
   Hor: P + 17 - 24 = 3×1.5
   P = 11.5 N
4. Parallel to surface:
   74 - 60 = 8a
   a = 14/8 = 1.75 m s^{-2}
1. Parallel to ground:
   Tcos30 - 40 = 8×3 T = 73.9 N

Page41

2.

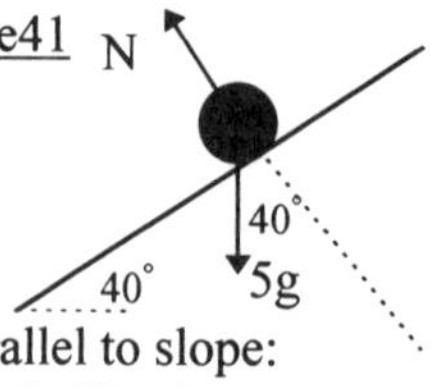

Parallel to slope:
5gsin40 = 5a a = 6.30 m s^{-2}

3.

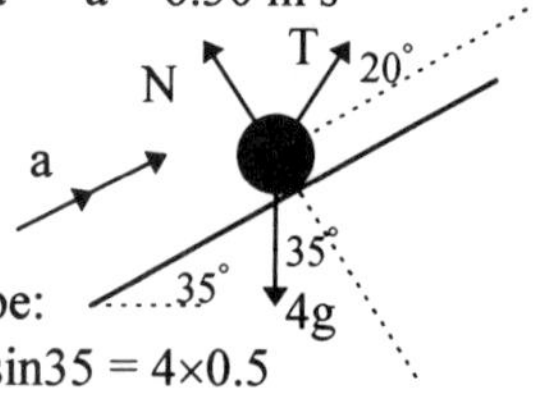

Parallel to slope:
Tcos20 - 4gsin35 = 4×0.5
T = 26.1 N

4.

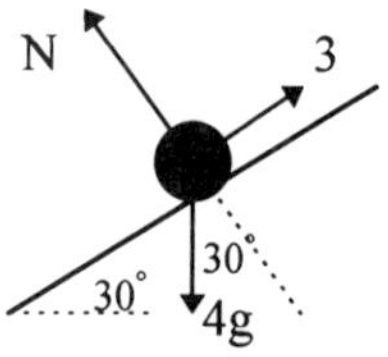

Parallel to slope:
4gsin30 - 3 = 4a a = 4.15 m s^{-2}

5.

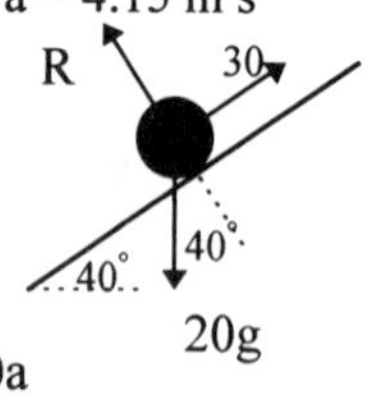

(a)Parallel to slope:
20gsin40 - 30 = 20a
a = 4.80 m s^{-2}
(b) '$v^2 = u^2 + 2as$'
= 0 + 2×4.799..×2
v = 4.38 m s^{-1}

Page42

1. For the 35kg sack:
   35g - T = 35a
   For the 25kg sack:
   T - 25g = 25a
   Add 10g = 60a
   a = g/6 = 1.63 m s^{-2}
   substitute T - 245 = 25 g/6
   T = 286 N

2. A: T = 3a
   B: 2g - T = 2a
   Add 2g = 5a
   $a = 2g/5 = 3.92\ m\ s^{-2}$
   Substitute T = 3×3.92 = 11.8 N

Pulley is smooth
A and B are particles
String is light and inextensible

Page43

3.(a) B: 11g -T = 11a
107.8 - T = 11a
A: T - 30 - 5gcos60 = 5a
T - 54.5 = 5a
Add 53.3 = 16a $a = 3.33\ m\ s^{-2}$
(b)T - 54.5 = 5×3.33125
T = 71.2 N
(c) '$s = ut + ½ at^2$'
= ½ ×3.33125×1 = 1.67 m

4. Whole system:
200-40-60 = 12a
100 = 12a $a = 8.33\ m\ s^{-2}$
block B: T - 40 = 5× 100/12
T = 81.7 N

Page44

5.(ii) $N_1 = m_1g$ $N_2 = m_2g$
(iii) $F = R_1 + R_2$
(iv) $F = R_1 + T$
(v) $T = R_2$
(vi) $F - R_1 - R_2 = (m_1 + m_2)a$
(vii) $F - R_1 - T = m_1a$
(viii) $T - R_2 = m_2a$

Page45

6. (i) A rope passing over asmooth pulley, has the same tension either side of the pulley, assuming the rope is light and inextensible.
(iii) Vert: 2Tsinθ = Mg
Hor: Tcosθ = Tcosθ not useful
(iv) No change in tension. If load remains in equilibrium T cannot change, so T in AB remains the same.
(v) x: 80cos40 - 80cos50
= 9.860546675
y: 80sin40 + 80sin50
= 112.7065642
Magnitude: √12800 = 113N
tanθ = 11.43 θ = 85° below BD

Page46

1. **v** = 10t - 4
   a = 10
2. $v = 5/2\ t^2 - 3t + c$
   v = 5 when t = 0 ∴ 5 = c
   $v = 5/2\ t^2 - 3t + 5$
   $s = 5/6\ t^3 - 3/2\ t^2 + 5t + c$
   s = 0 when t = 0 ∴ 0 = c
   $s = 5/6\ t^3 - 3/2\ t^2 + 5t$

Page47

3. $\mathbf{v} = (6t^2 + 6)\mathbf{i} + (2t - 4)\mathbf{j}$
   $\mathbf{a} = 12t\mathbf{i} + 2\mathbf{j}$
4. $\mathbf{v} = 5/2\ t^2\mathbf{i} - 3t\mathbf{j} + c$
   $5\mathbf{i} + \mathbf{j} = c$
   $\mathbf{v} = 5/2\ t^2\mathbf{i} - 3\mathbf{j} + 5\mathbf{i} + \mathbf{j}$
   $\mathbf{v} = (5/2\ t^2 + 5)\mathbf{i} + (-3t + 1)\mathbf{j}$
   $\mathbf{r} = (5/6\ t^3 + 5t)\mathbf{i} + (-3/2\ t^2 + t)\mathbf{j} + d$
   r = 3**i** when t = 0 ∴ 3**i** = d
   $\mathbf{r} = (5/6\ t^3 + 5t + 3)\mathbf{i} + (-3/2\ t^2 + t)\mathbf{j}$
5. **v** = 10t**i** + 8**j**
   when t = 3, **v** = 30**i** +8**j**
6. $\mathbf{v} = 6t^2\mathbf{i} + 3\mathbf{j}$
   when t = 1, **v** = 6**i** +3**j**
   Speed = $\sqrt{(36 + 9)} = 6.71\ m\ s^{-1}$
7. $\mathbf{v} = (3t^2 - 2t)\mathbf{i} + (-10t)\mathbf{j}$
   a = (6t - 2)**i** - 10**j**
8. **a** = 6t**i** **a** = 6**i**
9. $\mathbf{r} = 6t\mathbf{i} + t^2\mathbf{j} + c$
   r = 2**i** + 3**j** when t = 0 ∴ c = 2**i** + 3**j**
   $\mathbf{r} = (6t+2)\mathbf{i} + (t^2 + 3)\mathbf{j}$
   t = 2, **r** = 14**i** + 7**j**

Page48

10. $\mathbf{r} = (3/2)\ t^2\,\mathbf{i} + (-6/3)t^3\mathbf{j} + c$
    t = 0 2**i** + **j** = c
    $\mathbf{r} = (3/2\ t^2 + 2)\mathbf{i} + (-2t^3 + 1)\mathbf{j}$
    t = 2 **r** = 8**i** - 15 **j**
    Distance = √(64 + 225) = 17 m
11. $\mathbf{v} = t^2\mathbf{i} - t\mathbf{j} + c$
    t = 0 **r** = 4**i** + 5 **j**
    $\mathbf{v} = (t^2 + 4)\mathbf{i} + (-t + 5)\mathbf{j}$
    $\mathbf{r} = (t^3/3 + 4t)\mathbf{i} + (-t^2/2 + 5t)\mathbf{j} + c$
    r =0 when t = 0 ∴c = 0
    $\mathbf{r} = (t^3/3 + 4t)\mathbf{i} + (-t^2/2 + 5t)\mathbf{j}$
12. a = 4 - 2t
    0 = 4 - 2t at t = 2
    and $v = 8 - 4 = 4\ m\ s^{-1}$
    $s = 2t^2 - t^3/3 + c$
    s = 0 when t = 0 ∴ c = 0
    $s = 2t^2 - t^3/3$
    t = 3 s = 18 - 9 = 9 m
    s = 0 when $2t^2 - t^3/3 = 0$
    $t^2(6 - t) = 0$
    t = 0 (at start) or t = 6 seconds
13. $v = 20t - 3t^2 + c$
    v = 7 when t = 0 ∴ c =7
    $v = 20t - 3t^2 + 7$
    t = 7, v = 140 -147 + 7 = 0
    a = 20 - 6t = 0 at t = 10/3 secs.
    $v = 121/3\ m\ s^{-1}$
    $$s = \left[10t^2 - t^3 + 7t\right]_0^7 = 196\ m$$

Page49

14. $v = t^2 - 8t + c$
    t = 0 v = 12 ∴ c = 12
    $v = t^2 - 8t + 12$
    v = 0 = (t - 6)(t - 2)
    stops at t = 2 and t = 6 seconds.
15.(a) $s = 5t^2 - 2/3\ t^3 + c$
    t = 0 , s = 0 ∴ c = 0
    $s = 5t^2 - 2/3\ t^3$
    t = 3 s = 45 -18 = 27 m
(b) $s = 3t^2/2 + 3t + d$
    t = 3, s = 27   d = 9/2
    $s = 3t^2/2 + 3t + 9/2$
    t = 5 s = 57 m
16. a = 4 -2t
    v = 0 = (6 - t)(2 + t) when t = 6
    $a = 4 - 12 = -8\ m\ s^{-2}$
    a = 4 - 2t = 0 when t = 2
    $s = 12t + 2t^2 - t^3/3 + c$
    t = 0 , s = 0 ∴ c = 0
    $s = 12t + 2t^2 - t^3/3$
    t = 2 s = 24 + 8 - 8/3 = 88/3 m
    Particle stops stops at t = 6
    $$\text{Total distance} = \left|\int_0^6 v\,dt\right| + \left|\int_6^9 v\,dt\right|$$
    =72 + 45 = 117 m

Page50

17. $\mathbf{v} = (2t - 4)\mathbf{i} + (3t^2 + 2kt)\mathbf{j}$
    at rest 2t - 4 = 0 t = 2
    $3t^2 + 2kt = 12 + 4k = 0$ k = -3
    t = 2 **r** = -4**i** - 4**j**
    Distance √(16 + 16) = 5.66
18. $\mathbf{v} = (3t^2 - 3)\mathbf{i} + 8t\mathbf{j}$
    Parallel to **i** + **j** if coefficient of **i** and **j** are the same.
    So $3t^2 - 3 = 8t$
    $3t^2 - 8t - 3 = 0 = (3t + 1)(t - 3)$
    t = -1/3 or 3 so t = 3 seconds
19. **v** = 4**i** - (2t - 5)**j** + 12**k**
    t = 2 **v** = 4**i** + **j** + 12**k**
    a = -2**j** + 6t**k**
    t = 2 **a** = -2**j** + 12**k**

Page51

1. a = 6t**i** + 9**j**
   F = 5(6t**i** + 9**j**) = 30t**i** + 45**j**
   t = 1 **F** = 30**i** + 45**j**
2. **F** = 6t**i** - 9**j** = 3**a**
   a = 2t**i** - 3**j**
   $\mathbf{v} = t^2\mathbf{i} - 3t\mathbf{j} + c$
   t = 0 , 2**i** + 5**j** = c
   $\mathbf{v} = (t^2+2)\mathbf{i} + (5 - 3t)\mathbf{j}$
3. $\mathbf{v} = \frac{1}{12}\left\{(18t^2 - 5)\mathbf{i} + (4t^3 + 6t^2)\mathbf{j}\right\}$
   $\mathbf{a} = \frac{1}{12}\left\{(36t)\mathbf{i} + (12t^2 + 12t)\mathbf{j}\right\}$
   $= 3t\mathbf{i} + (t^2 + t)\mathbf{j}$
   $\mathbf{F} = 10\mathbf{a} = 30t\mathbf{i} + 10(t^2 + t)\mathbf{j}$
   t = 3 **F** = 90**i** + 120**j**

Magnitude 150 N
at 53.1° to **i** vector.

Page53

1. $U_H = 19.6\cos\alpha = 5.488$
   $U_V = 19.6\sin\alpha = 18.816$

(a)Vertically:
$U_V = 18.816$ $V_V = 0$ $a = -9.8$
Using '$v = u + at$'
0 =18.816 - 9.8t
t =1.92 sec

(b) Vertically:
$U_V = 18.816$ s = 0 $a = -9.8$
Using '$s = ut + \frac{1}{2}at^2$'
$0 = 18.816\,t - 4.9t^2$
=t(18.816- 4.9t)
t = 0 (at projection) or
t = 3.84 sec.
∴ time to hit ground is 3.84 sec.

(c) Horizontally:
$U_H = 5.488$ t = 3.84
Using '$s = ut$'
x = 5.488×3.84= 21.1.m

(d) Vertically:
$U_V = 18.816$ $V_V = 0$ $a = -9.8$
Using '$v^2 = u^2 + 2as$'
0 = 354.041856 - 19.6s
s = 18.06336= 18.1 m

(e) Speed of projectile at time t = 1.
Vertically:
$U_V = 18.816$ $a = -9.8$ $t = 1$
using ' v = u + at'
$V_V = 18.816 + -9.8\times1 = 9.016$
Horizontally: $V_H = U_H = 5.488$
$\therefore$ Speed = $\sqrt{111.406} = 10.6$ m s^{-1}

Page54

2. (i) Horizontally:
$x = 22\cos\theta\, t = 17.6t$
Vertically:
$y = 22\sin\theta\, t - 4.9t^2 = 13.2t - 4.9t^2$
(ii) $x/17.6 = t$,
$y = 13.2x/17.6 - 4.9x^2/17.6^2$
$y = 0.75x - 0.0158x^2$
(iii) $y = 2.7469 = 2.75$ m
(iv) $9 = 13.2x/17.6 - 4.9x^2/17.6^2$
$0.0158x^2 - 0.75x + 9 = 0$
$x = \dfrac{0.75 \pm \sqrt{(0.5625 - 0.56947314)}}{0.0316}$
Cannot calculate the square root, so no x coordinate for y = 9.
$\therefore$ projectile does not rise above 9m.

3. Vertical motion: $s = ut + \frac{1}{2}at^2$
$10 = 0 + \frac{1}{2}9.8 \times t^2 = 4.9t^2$
$\Rightarrow \dfrac{10}{4.9} = 2.0408163 = t^2$
t = 1.428571429
t = 1.43 secs (3 s.f.)
Horizontal motion: s = ut
$x = 14 \times t = 14 \times 1.428571429$
= 20.0 m

Page55

4. $v = 0 + 9.8 \times 0.5$
= 4.9
Speed = $\sqrt{(30^2 + 4.9^2)}$ 30.4 m s^{-1}
$\tan\theta = 4.9/30$ $\theta = 9.28^\circ$ below hor

5. Hor: $u_H = 15\cos60 = 7.5$
Vert: $u_V = 15\sin60 = 12.99038$
Vert: '$v^2 = u^2 + 2as$'
0 = 168.75 - 19.6s
s = 168.75/19.6 = 8.61 m

Page56

6. Taking up as +ve
(a) Vert: '$s = ut + \frac{1}{2}at^2$'
$-50 = 15t - 4.9t^2$
$4.9t^2 - 15t - 50 = 0$
t = 5.07 secs (-2.01 not pos.)
(b) Hor: 's = ut'
$x = 30\cos30\times5.072766$
= 132 m

7. (i) Hor: 's = ut'
40 = 20t t = 2 secs
Vert: '$s = ut + \frac{1}{2}at^2$'
$s = 40\sin60\times2 - 4.9\times4 = 49.7$
Clears mast by 19.7 m
(ii) Vert: '$v^2 = u^2 + 2as$'
0 = 1200 - 19.6s s = 61.2 m
(iii) Vert: '$s = ut + \frac{1}{2}at^2$'
$0 = 40\sin60\times t - 4.9t^2$
$= t(40\sin60 - 4.9t)$
t = 0 (at start) or t = 7.07 secs
(iv) Hor: 's = ut'
$x = 40\cos60\times7.069595$
= 141 m

Page57

8. Taking up as positive
(i) Vert: '$v^2 = u^2 + 2as$'
$0 = (25\sin30)^2 - 19.6s$
s = 7.97
Height above plane 37.97 = 38.0 m
(ii) Vert: '$s = ut + \frac{1}{2}at^2$'
$-30 = 12.5t - 4.9t^2$
$4.9t^2 - 12.5t - 30 = 0$
t = 4.06
(iii) Hor: 's = ut'
$x = 25\cos30\times4.05928$
= 87.9 m
(iv) Hor vel^y: 25cos30 = 21.6506
Vert vel^y : 'v = u + at'
$v = 12.5 - 9.8\times4.05928$
Speed = $\sqrt{1213} = 34.8$ m s^{-1}

9. Vertically: '$s = ut + \frac{1}{2}at^2$'
$2 = 0 + .5\times9.8\times t^2$
$t^2 = 0.40816 \Rightarrow t = 0.639$ secs
Horizontally: s = ut
$300 = u \times 0.638876565$
$\Rightarrow u = 470$ ms^{-1} (3 s.f.)

Page58

1.(i) 20 000-3 000-1 000-1 000
=60 000a
a = 0.25 m s^{-2}
(ii) T-1 000 = 15 000×0.25
T = 4750 N
(iii) $T_1 - 5750 = 15\,000\times0.25$
$T_1 = 9500$ N
(iv) Whole train -
(5000+2000+1000+1000) = 60000a
a = -9000/60000 = -0.15 m s^{-2}
Deceleration 0.15 m s^{-2}
Rear truck
$T_2 - 1000 = 15000\times(-0.15)$
$T_2 = -1250$ N i.e compression
Front truck
$T_1 - T_2 - 1000 = 15000\times(-0.15)$
$T_1 = -2500$ N i.e compression

Page59

2. (i) Up +ve
Vertically: Using '$v^2 = u^2 + 2as$'
$0 = u^2 - 19.6\times90$
$u^2 = 1746$
u = 41.8 m s^{-1}
(ii) Horizontally: Using 's = ut'
ucos60×2 = 40
u = 40 m s^{-1}
(iii) Vertically:
Using '$s = ut + ½at^2$'
h = 40sin60 - 4.9×4 = 49.7 m
(iv) Vertically:
Using '$s = ut + ½at^2$'
$30 = 50\sin60\, t - 4.9t^2$
$4.9t^2 - 43.30127t + 30 = 0$
t = 0.7578 or 8.079
above 30 m for 7.32 secs
(v) Vertically:
Using '$s = ut + ½at^2$'
$100 = 3\times50\sin\theta - 4.9\times9$
$\sin\theta = 0.96067$ $\theta = 73.9^\circ$

Page60

3. (i) 40 N
(ii) T - 40 = 80×0.3
T = 64 N
(iv) T = 40 + 80gsin30 = 432 N
(v) T - 40 - 80gsin30 = 80×(-0.2)
T = 416 N
(vi) 'v = u + at'
0 = u - 2
'$s = ut + ½at^2$'
s = 20 - 0.1×100 = 10 m
(or ' $s = vt - ½at^2$'
= 0 + 0.1×100 = 10 m)

Page61

4.(i) t = 0 **r** = 9**j** Ans 9 m
(ii) t = 3 **r** = (81/2 -27)**i** + (9 - 9)**j**
= 13.5**i** i.e. at (13.5,0)
(iii) **v** = $(9t - 3t^2)$**i** + (-2t)**j**
a = (9 - 6t)**i** + (-2)**j**
(iv) **v** = 6**i** - 2**j**
Speed = $\sqrt{(36 + 4)} = 6.32$ m s^{-1}
(v) **a** = -3**i** - 2**j**
Mag = $\sqrt{(9+4)} = 3.6056$
= 3.61 m s^{-2}
F = 20×3.605.. = 71.1 N
(vi) Horizontal velocity = 0 for 3t(3 - t) = 0 i.e. t = 0 or 3 at the beginning and end of the motion.

Page62

5. (ii) 50g in AB and BC
(iii) 500 -490 =50a
a = 0.2 m s^{-2}
(iv) Using '$v^2 = u^2 + 2as$'
1 = 0.4s s = 2.5 m
(v) Using '$s = ut + \frac{1}{2}at^2$'
$4 = -t + 4.9t^2$
$4.9t^2 - t - 4 = 0$
t = -0.807 or 1.01
$\therefore$ t = 1.01
(vi) Using '$s = ut + \frac{1}{2}at^2$' for 2 m
2 = 8a a = 0.25 m s^{-2}
Using '$v^2 = u^2 + 2as$' for all 5 m
4 = 10a a = 0.4 m s^{-2}
Not the same acceleration - so not a constant force.

Page63

6. (a) 4100-800-300 = 2000a
a = 1.5 m s^{-2}
(b) caravan: T - 300 = 500×1.5
T = 1050 N
(c) whole system:
P - 800-300-2000×9.8×1/10 =2000a
u = 10 v = 20 t = 16 ' v = u +at'
20 = 10+16a a = 0.625 m s^{-2}
P - 3100 - 1960 = 2000×0.625
P = 6310
(e) -(500g×1/10 + 300) =500a
a = -1.58 m s^{-2}
(f) 0 = 400 - 3.16s s = 127 m